Being Matt Murdock

ONE FAN'S JOURNEY INTO THE SCIENCE OF DAREDEVIL

Christine Hanefalk

Creator of *The Other Murdock Papers*

Being Matt Murdock: One Fan's Journey Into the Science of Daredevil by Christine Hanefalk

Cover design by Phyllis Sa, featuring art by Monique Müge

ISBN: 978-91-987965-0-6 (Paperback edition)

In Memoriam
Aaron Kimel and Van Diaz

CONTENTS

NOTES FROM THE AUTHOR

This book has a companion website at www.scienceofdaredevil.com. This is where you will find bonus content and complementary information, including panels from the comics to go with the descriptions in the text. It has been important to me to write the kind of book that Matt Murdock himself would be able to enjoy, which is why it's never assumed that the reader needs to be able to see the art I'm talking about, and there are no pictures. However, for those of you who want to see for yourselves what I'm attempting to describe, the companion website is where you will find all that good stuff, sorted by chapter.

An additional thing to keep in mind has to do with the convention of how the comic books are referenced as sources. Like many other comic book titles, *Daredevil* has gone through a series of different volumes. Each volume starts the book with a new #1, which can be confusing for anyone who hasn't followed the various iterations of *Daredevil* very closely. There is also the additional complication of the book occasionally going back to the original numbering of the first volume, particularly around major anniversaries or milestones. This is why there is a *Daredevil* #500, but no *Daredevil* #400.

Typically, runs are identified by putting the publication year of the first issue of the particular run in parentheses when a specific issue is mentioned. For Daredevil, volume 1, this would be 1964. Volume 2 launched in 1998, and volume 3 in 2011. There have been additional volumes in the years since, but most of the examples I cover in this book are from these first three. To not make the page too cluttered, I've been selective about when to indicate the particular volume, and have done so mostly when it's not clear from the context or when there is a need to indicate a switch between volumes.

INTRODUCTION: SCIENCE IS YOUR FRIEND

The book you are about to read could jokingly be described as a work of "fan non-fiction." Before you think that this sounds like the words of a writer selling herself short, or some kind of indictment of the level of ambition behind this project, let me assure you that this could not be further from the truth.

First of all, many fan fiction writers are good writers in their own right, with a firm grasp of the fictional worlds they explore. Besides, as a friend and fellow Daredevil fan liked to point out, the stories in the comics are themselves merely a form of officially sanctioned fan fiction. The characters and the worlds they inhabit have been passed from one creative team to the next for decades, with successive writers and artists putting their own twists on the stories they tell, adding new elements, and reshuffling old ones. And, I would certainly hope that Daredevil's creators consider themselves fans of the character as well.

Secondly, the fact that I come to this work as a Daredevil fan has been vital. Many of the questions I have sought to address here first popped into my head many years ago when I first discovered Daredevil for myself, such as:

- "What *is* Daredevil's radar sense, really?"
- "If you had 'super hearing,' what would you really be able to hear, and from what distance?"
- "What might you be able to find out about people and places through scent alone?"

It's true that you need to know a thing or two about science to write a book that attempts to answer, or at least explore, these kinds of questions. You also need to be a massive geek. It is a badge I claim with pride.

If you have picked up this book primarily as a reader of science non-fiction and are not already intimately acquainted with Daredevil, let me offer a brief introduction to the character. The rest of you can feel free to skip ahead!

Daredevil is a Marvel comic book superhero created during the so-called Silver Age of comics. Introduced to the world in the pages of *Daredevil* #1 (1964) by writer Stan Lee and artist Bill Everett, our hero represented a twist on an established pattern.[1] Regular Marvel readers had already witnessed the consequences induced by the bite of a radioactive spider (*The Amazing Spider-Man*), exposure to cosmic rays (*The Fantastic Four*), or high levels of gamma radiation (*The Incredible Hulk*).[2]

The young Matt Murdock, not yet a superhero, would also find himself transformed by a chance encounter with radiation, struck across the eyes by radioactive waste after pushing an old man out of the path of a speeding truck. The accident instantly destroyed Matt's eyesight, but he came to realize that his remaining senses had been heightened to a remarkable degree! He had even developed a sort of "radar sense" that allowed him to move safely through a world he could no longer see.

Matt had been raised by a loving but strict single father, the struggling boxer "Battling'" Jack Murdock. Jack had always encouraged the boy to study hard to make something of himself and forbidden him from using his fists. Being a good student who stayed away from fights and rough play made Matt the target of school bullies who called him Daredevil, sarcastically noting that he was anything but. Following his accident, Matt began to work out in secret, using his new heightened abilities to perform physical feats that now came easily to him. At the same time, he would do as his father wished and study hard, eventually going to college and becoming a lawyer.

Not long before Matt's graduation, tragedy struck again. Jack had become mixed up with the mob, specifically a shady figure called The Fixer who, as the name suggests, made his money fixing boxing matches. With his son in the audience one night for the fight of his career, and wanting nothing more than to make him proud, Jack decided to go against the Fixer's order to take a dive. He wound up paying for this decision with his life.

Matt would go on to start the law firm Nelson & Murdock with his best friend and former college roommate Franklin "Foggy" Nelson. However, frustrated that his father's killers had not yet been brought to justice, Matt decided that he needed to go beyond the confines of the law. By reclaiming the name the school bullies had used to taunt him and crafting a costume to hide his identity, he took up his fight for justice as Daredevil, the Man Without Fear! (Using highly questionable reasoning, Matt also decided that creating this separate persona had the added advantage of releasing him from the promise he had made to his father to never resort to physical violence.)

Over the many decades since his first appearance, Daredevil has been featured in a wide range of tonally quite different types of stories. However, many would point to the comic book run by writer and artist Frank Miller in the early 1980s as setting the standard for the modern understanding of the character. Miller pitted Daredevil against the Spider-Man villain Wilson Fisk, better known as the Kingpin, and elevated the psychopathic killer Bullseye to the status of archenemy. The mysterious Elektra was introduced to Matt's past and present as the old college girlfriend turned assassin.

Another newcomer to the world of Daredevil was Stick, the blind old man who was revealed to have been young Matt's guru. The stories featured both mystical elements like ninjas, and organized crime and corruption. Miller's take on the character didn't prevent later writers and artists from putting their own mark on *Daredevil*, but many of the classic elements we now associate with the title character came out of this era.

Today, most readers view Daredevil as a prime example of a "street-level" hero, one who focuses more energy and attention on neighborhood crime and corruption than battling aliens from outer space. This also provides a good thematic fit for his legal career as Matt Murdock, and he is often seen using both guises to tackle injustice. Daredevil is also very closely associated with the New York neighborhood of Hell's Kitchen and feels a great sense of duty to its inhabitants.

Daredevil will forever be a character defined by contradictions. He is a lawyer who breaks the law and a blind man with heightened senses that subvert the readers' expectations. He is equipped with enough moral ambiguity to satisfy everyone's need for depth and complexity. For me personally, Daredevil will always stand out as an amazingly interesting character, and I would assume that the same is true for anyone who has come under the spell of Matt Murdock.

Speaking of the civilian persona, it is not a coincidence that I have chosen to give this book the title *Being Matt Murdock*. While Marvel has long excelled at featuring the people behind the masks as prominently as their costumed alter egos, I cannot help but feel that this is especially true for Matt Murdock. His civilian persona has been at the center of some of the most interesting *Daredevil* stories ever told, and his unique combination of sensory enhancements and deficits is guaranteed to affect every aspect of Matt Murdock's life, whether he is in costume or not.

During my many years of blogging at *The Other Murdock Papers*, a blog devoted specifically to Daredevil that I launched in 2007, I have been fortunate enough to connect with other fans who share much of my appreciation for the character, as well as my occasional frustrations. When I gradually turned to writing more frequently on the topic of Daredevil's senses and how they relate to real-world science, I was worried that it might turn some people off. Perhaps the subject would be too technical, or my takes too... well, "nit-picky" is a word that comes to mind. I was pleasantly surprised to discover that my science posts consistently ranked among the most popular – and the most commented on.

I think this has been partly due to the lack of information on this topic elsewhere. Aside from the occasional mention in the popular science press, usually when a writer wants to make a culturally relevant point about echolocation, there has been little written about Daredevil from this perspective. I also like to think that my regular readers have enjoyed my science posts because they have conveyed at least some of the enjoyment of writing them. If I'm having fun, the odds are good that my readers are too. I have brought that same passion, and more than occasional irreverence, to the writing of this book.

However, I am writing this book not only because I enjoy the process of researching obscure topics and picking apart throw-away lines from decades-old comics. While all of that is fun and engaging, there is also an important message at the heart of this project. As great of a character as Daredevil is, I'm convinced that it's possible to come to a fuller understanding of his heightened senses, and the implications of the one he's missing. In my mind, taking scientific realities into account to a greater degree than has been the case historically would not pose a threat to any of the things fans love about Daredevil, but might hold the key to more consistent and engaging depictions of Matt Murdock's world. In that sense, this book is also intended as a measured critique of how the comics have delivered on the deeply fascinating concepts at the heart of this character.

I suspect that a few of you are probably thinking that holding comic book characters to *any* scientific standard will ruin the whole premise. At the same time, and somewhat paradoxically, many comic book fans care a great deal about rules. Most readers would raise a concerned eyebrow if Captain America suddenly took to the sky flying or if the Hulk opened an inter-dimensional portal. While we allow for the existence of both of these abilities in the Marvel Universe, or the superhero genre more broadly, fans expect a sort of framework for which character can do what and under what circumstances. I would argue that whether people are aware of it or not, scientific reasoning often comes into play when readers or viewers react to something that seems to break the pre-established rules.

Consider, for example, the scene depicting Steve "Captain America" Rogers, played by actor Chris Evans, taking down a helicopter in the 2016 movie *Captain America: Civil War*. When the scene begins, we see Rogers run onto the helicopter pad and spot his friend (and current adversary) Bucky Barnes taking off. To stop him, Rogers jumps off the ground to grab the landing gear and somehow *manages to pull the helicopter back down*. After a hop and a skip, he next takes hold of a metal bar that runs along the edge of the helicopter pad, and we are treated to a few seconds of Chris Evan's bulging biceps before Bucky, played by Sebastian Stan, goes attack helicopter on his old friend.

If you are anything like me, you probably chuckled a bit at the first part of the scene where Steve Rogers has to rely on *nothing but his own body weight*. At the same time, you were probably okay with the latter part of the scene, where he keeps the helicopter from flying away by holding on to the railing. Whether you can put your finger on it or not, the difference between these two conditions is that the second follows naturally from the "rules" that govern Captain America's superpower – which happens to be his significantly enhanced physical strength – whereas the first does not.

In his excellent book, *The Physics of Superheroes*, physicist and writer James Kakalios introduces the concept of the "miracle exception," which is one I find extremely helpful when talking about these things. In Kakalios's own words:

"Of course, nearly without exception, the use of superpowers themselves involves direct violations of the known laws of physics, requiring a deliberate and willful suspension of disbelief. However, many comics needed only a single 'miracle exception' – one extraordinary thing you have to buy into – and the rest that follows as the hero and villain square off would be consistent with the principles of science."

Looking at a character like Daredevil, you have someone who – like Captain America, Spider-Man, the Hulk, and many others – can easily be conceptualized as a "miracle exception" character. This doesn't mean that there isn't room for characters that annihilate the laws of physics, but rather that such characters might best be described and understood as purveyors of magic or something close to it.

Buying into the idea that Matt Murdock has heightened senses doesn't mean that we're not allowed to question how these senses are used. With this in mind, it shouldn't be a surprise that scenes that read more like magic than heightened senses usually entail a fundamental misunderstanding of how the different forms of energy that our bodies are sensitive to actually work, both on a physical level and in their interactions with sensory systems. They suffer from what I have taken to calling "stimulus problems." (We'll dive into stimulus problems in chapter two.) It is not at the point of the "miracle exception" that things fall apart.

I firmly believe that if the creators in collective control of Daredevil's future adventures were to learn more about the science that applies to a character like this one, it would benefit the storytelling, make the "rule book" more robust, and the interpretation of his powers more consistent. And they wouldn't have to come up with ridiculous ways to get him into trouble either. There is nothing to fear. Science is your friend.

When we focus on Daredevil, in particular, there are two additional reasons to pay more attention to the science, and they are perhaps even more compelling than the one I've touched on already.

Let's begin with the fact that the specifics of Matt Murdock's particular superpowers are fascinating. More so than Steve Roger's strength and even Peter Parker's sticky fingers (I'm not even going to attempt an explanation of the "spider-sense"). How and why sensation works in the first place, here in the real world, is a fascinating topic. The world we occupy – with its sights, sounds, scents, and more – is brought to life by our brains, via the specialized connections our nervous system makes with the outside world. Some of the molecular micro-machinery that facilitates sensation is as old as life itself.

That we are able to sense and perceive our surroundings is the most viscerally real aspect of consciousness and has inspired much philosophical thought. Our senses underpin *everything* we do; they connect us to other people, keep us safe and fed, and provide us with the means for both enjoyment and suffering. My love of Daredevil as a character is entangled with my lifelong interest in this particular corner of the physical and life sciences. Exploring the science of Daredevil can help us understand ourselves and better appreciate the processes that bring our world to life for us.

The real science of sensation and perception can also foster a better understanding of Matt Murdock. While we don't necessarily need to account for the nature of his abilities – his "miracle exception" – I will still attempt to do precisely that, when possible. What is it that limits how and what we can sense? How would you theoretically go about tampering with those limits, and what would you be able to do if you could? How would such tampering help explain some of the things Daredevil can do? And what would it actually be like to *be* Matt Murdock?

The other reason scientific exploration seems particularly worthwhile in the case of Matt Murdock has to do with the fact that the accident which set him on the path to becoming Daredevil didn't only bring gifts. As a blind superhero, he is one of the few fictional characters who can legitimately be described as having both a disability and "superabilities." I would argue that this dichotomy remains under-explored. Instead, Matt Murdock's heightened senses are often presumed to mostly cancel out his blindness, making it seem more like a façade than something real and consequential. I believe that this stems from both the wishful thinking that one might argue has been built into the character from the outset, in the form of "compensation," as well as a misunderstanding of the science.

The more heavily creators of *Daredevil* comic books and live-action projects alike lean into the need for the heightened senses to compensate fully, even when it makes little sense, the more inflated and needlessly otherworldly they become, at the risk of robbing the character of one of his most unique and interesting qualities. In this case, expanding into magic territory directly threatens to undermine the authenticity of a core characteristic. That so many fans appear to have the impression that Daredevil can basically "see" should not be taken as a testament to his impressive power set but as a failure on behalf of often very gifted storytellers to communicate the totality and complexity of his unique perspective.

In reality, the functional implications of Matt Murdock's blindness, combined with his heightened senses, would depend entirely on the context and situation he finds himself in, which is not unlike disability in the real world. By applying basic scientific reasoning to all facets of Daredevil's sensory world, we end up with an understanding of this character that can account for both his heightened senses and his blindness and do so in authentic ways.

This book is divided into three parts. The first, *Foundations*, will cover some basic principles, beginning outside the realm of the natural sciences. As a blind superhero, Daredevil borrows heavily from common (mis)conceptions and myths about the blind. In that sense, he is not only the supposed product of radioactivity but of a rich and ancient literary tradition. This is the subject of the first chapter, which also takes a look at other fictional blind characters from within and beyond the comic book genre.

Chapters two through four take us deep into the scientific domain of the senses and brain more generally. We begin with a discussion of the importance of the stimulus itself and how our bodies – starting on a *molecular* level – interact with a diverse range of energy and matter. We will also look at what the brain does with the information that is passed on to it from the sensory organs, and how it adapts to new and unusual challenges, as it would have to in the case of the young Matt Murdock. With these chapters, I aim to take most of the magic out of sensation and perception and replace it with a scientifically sounder sense of awe.

In the second part of the book, *Super Senses*, we take a look at each of Daredevil's heightened senses and compare what we find with scenes and concepts from the comics and beyond. Starting with hearing in chapter five, we will look more closely at the physics of sound, and the biology of the mammalian ear. We tackle questions like "How far can you really hear?" and "How would you hear heartbeats?" In chapter six, we look at the special case of echolocation which, amazingly, allows even ordinary humans to use the world of sound in fascinating ways to detect silent objects. Chapters seven and eight, respectively, are devoted to Matt Murdock's more underutilized and under-appreciated senses, those of smell and touch.

The final part, *Radar Sense*, is devoted entirely to exploring Daredevil's most enigmatic sense. Or should I say "senses"? Or *ability*? The fact is that trying to nail down the nature of the radar – that manifestation of his gifts that affords him the ability to "see" (sort of) – is not as straightforward as you might think. Chapter nine focuses on the surprising – and surprisingly *messy* – early history of the radar sense, while chapter ten represents my attempt to nail down its nature(s), or at least some common themes. Chapter eleven focuses on how the creators have portrayed the *experience* of the radar sense, with a special look at how the artists have attempted to render it on the page. In the final chapter, we look beyond the reaches of the radar sense to a study of how the impact of Matt Murdock's blindness has been understood – and sometimes forgotten – over the years.

I want you to walk away from reading this book with a heightened sense of awareness (see what I did there?) about what the senses – including your own! – actually do, and how. Going forward, you may even notice yourself reacting to things in the comics that you wouldn't have previously. Ideally, as one fan writing to another, I hope this book only deepens your love of Daredevil.

PART ONE
FOUNDATIONS

CHAPTER 1
THE LITERARY ADVENTURES OF THE SUPER-BLIND

 "Their senses had become marvellously acute; they could hear and judge the slightest gesture of a man a dozen paces away — could hear the very beating of his heart. Intonation had long replaced expression with them, and touches gesture, and their work with hoe and spade and fork was as free and confident as garden work can be. Their sense of smell was extraordinarily fine; they could distinguish individual differences as readily as a dog can, and they went about the tending of the llamas, who lived among the rocks above and came to the wall for food and shelter, with ease and confidence."

The Country of the Blind, by H. G. Wells (1904)

As Daredevil fans, we may prefer to think of our hero as being truly one of a kind. However, if we focus narrowly on Daredevil's basic premise as a superhero – that of a blind man whose remaining senses have become exceptionally acute to compensate for the loss of his sight – we soon come to realize that Daredevil was not created in a vacuum. Popular myths about the blind have been with us for centuries, and don't appear to be limited to any specific culture. Even those aspects of the character that might at first seem to be unique to the "Man Without Fear" have been (and remain) common in depictions of the blind in art and literature.

These patterns have not gone unnoticed by the highly entertaining wiki-style pop culture website TV Tropes, which lists several of the narrative devices or conventions relevant to fictional blind characters, and the topic of disability more broadly. Not surprisingly, Daredevil is found in several of the categories provided, suggesting that he is perhaps not unique so much as uniquely predictable. Ouch! Among the more obvious is "Disability Superpower," described in an entry that begins:

"A character is born with or acquires some handicap that prevents them from functioning normally. However, due to phlebotinum exposure or training, never mind Disability Immunity, they develop something that not only makes up for what's missing, but goes beyond it.

Blindness seems to be a popular one for this. Indeed, the entire trope seems to be based around the idea that blind people's other senses become more acute to compensate."

The mentioned "Disability Immunity" is further described as the sub-trope according to which "having a disability makes you immune to stuff that affects people who don't have the disability." Considering the many times Daredevil has been able to withstand everything from hypnotism to a frankly implausible number of blinding rays and similar devices, there's another box for us to check. Then there's the sub-trope labeled "Disability-Negating Superpower," for which our own Matt Murdock is also listed, along with the painfully obvious "Super Senses." If Matt himself were around to read this, he'd have to raise his hand and declare himself "guilty as charged."

For a less irreverent and more academic look at the same subject, particularly as it pertains to blindness, we can also turn to *Beneficial*

Blindness: Literary Representation and the So-Called Positive Stereotyping of People with Impaired Vision, a 2006 paper by David Bolt published in the *Journal of Disability Studies.*[1] Here, the author takes a closer look at this particular kind of representation of blindness in several works of fiction published since the late 19th century.

For anyone with more than a passing familiarity with Daredevil, a reading of the examples of the so-called positive stereotypes provided by Bolt is bound to inspire feelings of *deja-vu*. The first to hit too close to home revolves not around heightened senses *per se*, but the idea that blindness would enable a person to pick up certain skills more readily, or take more easily to intellectual pursuits. One such example comes to us in *The Gift of Sight*, an 1898 short story by Indian writer Rabindranath Tagore. When the character Kusum is blinded, any concerns about her future as a housewife are quickly put to rest:

"Within a short time I had learnt to carry out my customary tasks through sound and smell and touch. I even managed much of the housework with greater skill than formerly."

From a rational point of view, such enhanced skills make very little sense. With training and experience, blind people are perfectly capable of performing most of the tasks we associate with good housekeeping. However, it is a bit of a stretch to suggest that the loss of vision would be a net *benefit* in this regard.

Bolt notes that the reasoning behind the stereotype that many tasks – even when they involve a significant visual component – can be performed with greater skill by the blind often hails from the notion that lacking visual distractions would enhance the ability to concentrate. Along these lines, we are offered another example from André Gide's 1919 novella *La Symphonie Pastorale* which tells us that the blind character Gertrude "showed more sense and judgment than the generality of young girls, distracted as they are by the outside world and prevented from giving their best attention by a multitude of futile preoccupations."

If we look at *Daredevil*, we'll note that the early creative teams in particular would have us believe that there was virtually no feat Matt Murdock could not perform with greater speed or skill (often both) than a sighted man, whether such enhancements could be expected to follow naturally from heightened senses or not. The very first issue of the comic even makes the dubious claim that Matt's accident had made him a better student!

For an example of the idea that blindness would relieve a person, namely Matt Murdock, of the distraction of vision, we need look no further than the first season of the Daredevil television show which debuted in 2015. In episode ten, *"Nelson vs Murdock,"* Matt tells his best friend Foggy about his old mentor Stick. In Matt's own words, Stick had taught him that "my blindness wasn't a disability, that sight was a distraction." To be fair, three episodes earlier, in *"Stick,"* we also learned that the very same mentor considers women, apartments, and furniture(!) to be distractions as well, so make of this particular statement what you will.

THE BLIND SUPER-SENSER

One might argue that the difference between Daredevil and Tagore's Kusum and other similar characters lies in the fact that the former is meant to have actual and explicitly defined *superpowers*. However, even the enhanced senses that are at the heart of how Daredevil operates have not historically been limited to the realm of the supernatural, or the world of superheroes, but are found among Bolt's examples of positive stereotypes.

Although the *Daredevil* comics never suggest that Matt Murdock would have acquired these heightened abilities without that helpful dose of radiation, it is interesting to note just how common heightened senses similar to Daredevil's have been in fictional depictions of blindness even when the character in question is presumably "just blind," without the benefit of Silver Age superhero magic.[2] The opening quote of this chapter, from H. G. Wells's *The Country of the Blind*, mentions such Daredevil staples as hearing heartbeats and being able to tell individuals apart by smell. Another one of Bolt's examples, so extreme that it would put even Matt Murdock to shame, comes from references to the character Andreas in Arthur Conan Doyle's 1906 novel *Sir Nigel*:

"...for it often happens that when a man has lost a sense the good God will strengthen those that remain. Hence it is that Andreas has such ears that he can hear the sap in the trees or the cheep of the mouse in its burrow. He has come to help us to find the tunnel."

On the topic of extreme hearing, we also have the case of the blind man in August Strindberg's *A Dream Play* (1901) who is able to hear his son across the sea, making Daredevil's ability to sift through the din of Hell's Kitchen appear modest by comparison:

"Now I hear the cable screech, and – something flutters and swishes like clothes drying on a line – wet handkerchieves, perhaps – and I hear how it snuffles and sobs, like people crying – perhaps the small waves lapping against the nets, or is it the girls on the shore."

That Daredevil draws inspiration from the same underlying myths as these other characters, even with the radiation serving as a convenient transformative agent, is further revealed by Matt's reaction to secretary turned love-interest Karen Page's insistence that he meet with an eye surgeon to investigate whether something can be done about his damaged eyes. Rather than balk at Karen's meddling attempts to "fix" him, he worries about what having his sight restored might do to his remaining senses. In *Daredevil* #9 (1964), Matt thinks to himself:

"If only I didn't fear that I would lose my super-senses if my vision returned! How ironic that *Daredevil*, the man without fear, is mortally afraid of ever regaining his sight!"

As far as Matt is concerned, his heightened senses are inextricably linked to his blindness, and he must remain blind in order to keep his other senses working at peak capacity. At this point, you might be asking: Is there no truth to the idea that losing a sense will make the others sharper? And, if so, wouldn't it be reasonable for Matt to assume that regaining his sight would cause his heightened senses to lose their edge?

We will have reason to return to this fascinating topic in more detail in chapter four, but suffice it to say that the literary and popular myths surrounding this idea are a gross exaggeration of a real, but much more modest, phenomenon. Blind people often learn to make better *use* of their remaining senses, and studies confirm that brain areas normally devoted to vision can find new and interesting applications. There are areas where the blind, as a group, outperform the sighted. So-called sensory compensation is real, and neural plasticity is a basic human trait – or else we would never be able to learn new skills – but it boils down to what the *brain* is doing. Even the most proficient of blind echolocators will perform normally on a standard hearing test. They will not be able to hear a heartbeat or the mice in their burrows. Sensory compensation alone does not a Daredevil make.[3]

THE BLIND MARTIAL ARTIST

We have established that Daredevil is far from the only fictional blind character imbued with extraordinary senses. Nor, as we will see, is he the only one to have put his skills to such hands-on pursuits as fighting supervillains in the streets. In fact, martial arts and related disciplines appear to be highly popular among the fictional "super-blind."

In the television show *Daredevil*, when Foggy learns that Matt was trained to be a fighter by an old blind man, he reacts predictably: "A blind, old man taught you the ancient ways of martial arts? Isn't that the plot to *Kung Fu*?" The character Foggy is referring to here is the blind Master Po, played by Keye Luke, who appears in both the original *Kung Fu* television series (1972-75), and *Kung Fu: The Movie* (1986).

Not at all surprisingly, Master Po is yet another character who, aside from wisdom and fighting skills, appears to be equipped with heightened senses. And again, there is no apparent need beyond Po's blindness to explain how this could be. In the pilot episode of *Kung Fu*, after having defeated the young Caine, Master Po instructs the boy to listen and asks whether he can hear the sound of his own beating heart or the grasshopper at his feet. When asked "Old man, how is it that you hear these things?," Po simply replies "Young man, how is it that you do not?"

Another well-known blind character who seems to be cut from the same cloth as Master Po is Zatoichi, the hero of numerous Japanese television movies, the first of which was released in 1962. Actor Shintarô Katsu would go on to play the character in no fewer than twenty-six movies, and a television series (1974-1979) spanning one hundred episodes. A more recent, high-budget, remake came out in 2003, starring Takeshi Kitano as the titular character.

Blinded as a young child, Zatoichi is a masseur and (secretly) a skilled swordsman. As he roams the countryside of Edo-era Japan, he frequently finds himself in situations where he must protect the innocent, or defend himself from attackers, while seeking redemption for his past sins as a yakuza. And just like Master Po, he finds frequent opportunities to astound the people around him with his acute senses, for instance while engaging in one favorite pastime: gambling.

Zatoichi predates *Daredevil*. By early 1964, the first five movies of the successful Japanese franchise had already been released. To my knowledge, Zatoichi has never been referenced as an inspiration for Daredevil, however. Though obvious similarities exist, they probably have more to do with the common well of literary blindness tropes from which both characters draw inspiration.

The likes of Zatoichi and Master Po show more obvious similarities to young Matt's master Stick, introduced in 1981 by Frank Miller in the pages of *Daredevil* #176, than they do with Daredevil himself. The early comics are not characterized by any obvious East Asian influence, and the main character's fighting style, as depicted during *Daredevil's* Silver Age, appears to have more in common with the standard superhero acrobatics of the time than with any particular martial arts discipline.[4]

When Frank Miller came along, however, Daredevil's world was quickly invaded by ninjas, along with other elements straight out of the American popular mash-up of martial arts and Asian legend that had been gaining in popularity since the late 60s. Miller himself, only twenty-one years old when he started as the penciler on *Daredevil* in late 1978, made no secret of his interest in the genre.[5] Outside of the *Daredevil* comic, the kung-fu craze had already sparked the creation of characters like Shang-Chi.[6]

Stick, and Master Izo – a character created much later, in 2008, by the *Daredevil* creative team of Ed Brubaker and Michael Lark – occupy a more mystical realm than Daredevil does. Stick, the leader of a benevolent band of martial artists known as the Chaste, was born blind and presumably drew all of his heightened abilities from his training and whatever vaguely defined supernatural sources his ninja order has access to.

There is about as little explanation offered for how Stick can do what he does, as there is in the case of Zatoichi and Master Po, though it is clear that there is a difference between Stick and his adventitiously blinded young recruit. While Stick seems to be of the opinion that Daredevil's abilities build on something which lies dormant in all of us, he acknowledges that "helpful" dose of radiation that Matt received as a child, this in a storyline we'll return to in chapter ten.

If Stick seems mysterious, Master Izo still has him beat. Izo is apparently hundreds of years old and founded the Chaste – the same group Stick would later come to lead – after leaving the now villainous order of the Hand. After he found the Hand straying from the right path, he willingly blinded himself by staring into the bright sun. This particular scene, from *Daredevil* #500, at first seems reminiscent of another common trope that conceptualizes blindness as punishment. The idea goes back to ancient literary traditions, with the well-known Greek myth of *Oedipus* being a particularly notable example.

Izo, too, points to his own weakness, and how he didn't wish to see what the Hand had become. However, the true reason for why he chose to sacrifice his sight turns out to have been less about punishment and more about wishing to achieve the kind of insight we might associate with someone like the ancient seer Tiresias: "No, I blinded myself so I could see this world better than all of you..." This draws on an entirely different blindness trope which juxtaposes physical blindness with a kind of spiritual enlightenment.

The addition of Stick and Izo to the Daredevil mythos suggests a new way of looking at the character and his origins. Regardless of how successive comic book creators have wished for us to understand Matt Murdock's heightened abilities, his basic premise rests heavily on common myths associated with the supposed compensations of blindness. After the introduction of Stick, any reader so inclined could look at Daredevil's story more as a variation on the classic tales of blind martial artists specifically.

It is also worth noting that the larger class of fictional blind warriors and martial artists is still being added to, far beyond the confines of the *Daredevil* comic. We see recent examples in Chirrut Îmwe, who was introduced to the *Star Wars* universe in the 2016 movie *Rogue One*. Chirrut, played by Donnie Yen, is a "warrior monk" type of character who has no problems taking out Storm Troopers, despite his blindness. (On the other hand, Storm Troopers collectively seem like the most inept military force in the galaxy.)

I would also like to mention the Apple TV+ show *See*, which premiered in 2019. *See* is set in a dystopian future where humans have lost the ability to see, except for a precious few that herald the return of sight. Despite some of the inevitable absurdities of the premise, I have to admit to being a fan of this show. Of course, with (nearly) every character being blind, *See* takes us far beyond the warrior archetype and into every niche of human enterprise. In fact, one of the more sympathetic aspects of *See* is that it normalizes blindness to such a degree that this is never what sets characters apart, allowing them instead to display every virtue and vice imaginable. They are neither pitiable nor inherently heroic.

THE BLIND SUPERHERO

As we've seen, Matt Murdock is not the only blind character with enhanced abilities, even in his own comic book. And while Daredevil is by far the best-known blind superhero today, particularly after the success of his recent television show and return to the big screen, he is not the only one to fit that description either. He wasn't even the first. That distinct honor goes to the DC Comics character Doctor Mid-Nite who was created in 1941.

The original Doctor Mid-Nite, known in his civilian life as Charles McNider, made his first appearance in *All-American Comics* #25 (1939), written by Charles Reizenstein, with art by Stanley Josephs Aschmeier. McNider is introduced as a physician and researcher who is called on by the police to treat a badly injured mob informant. While McNider is tending to his patient, a gangster affiliated with the local mob boss appears, and throws a grenade through the window, killing the man Dr. McNider had just miraculously saved and permanently blinding the doctor himself.

This chain of events takes obvious inspiration from the story of the Black Bat, a character who appeared in a fairly successful series of pulp magazine stories, the first of which was published in 1939.[7] The Black Bat is the moniker of District Attorney Tony Quinn who, like McNider, dares to go head to head with the mob, and is blinded in an acid attack in the courtroom. After losing his sight, Quinn spends a year barely leaving the house. He is finally visited by the mysterious Carol Baldwin who makes arrangements for him to have a corneal transplant, an operation which not only restores his sight but also allows him to see in the dark! Since the operation was carried out in secret, Quinn is able to use the public knowledge of his blindness to create the perfect secret identity as he turns to crime-fighting on the side.

Unlike Tony Quinn, Dr. McNider takes his blinding injury in relative stride. He decides to continue his research, with the help of his assistant Myra Mason, and to start a new career as the writer of true crime detective stories. Some time later, an owl crashes through

McNider's window during a thunderstorm, lands on his lap and accidentally loosens the bandages around his eyes. McNider now makes the startling discovery that he can see in the dark (which he somewhat perplexingly still recognizes as such). Again unlike Tony Quinn, he doesn't regain his vision fully, but is blind again when the lights are turned on. The creators provide an interesting explanation for how this can be:

"Explanation of Dr. McNider's strange phenomenon! Everyone knows an owl can see only at night... The iris of its eye cannot close off light to the proper degree and so the owl is blinded by daylight... The accident made Dr. McNider's eyes the same as the owl's... he can see in the dark – but light blinds him!!"

In reality, this delightful explanation is complete nonsense, and the supposed "everyone" mentioned above is sadly mistaken. Most owl species are nocturnal and have evolved an impressive visual system that can support such a lifestyle, but they are not blind during the day. In fact, none of the animals known for their ability to see well in low light conditions – which famously includes cats – can see anything in *complete darkness*.

Blissfully unaware of the logical challenges to his new abilities, McNider decides to keep the owl as a pet – naming it Hooty – and later takes it with him while fighting crime. First though, our hero needs to devise a costume and a set of goggles that will allow him to see during the day. The lenses he creates are described as "infrared," which presumably means that they shield him from visible light, but let the longer-wavelength infrared rays through. The idea seems to be that owls can see infrared light and that because Doctor Mid-Nite now has eyes that work like those of an owl, he can too.

Before we completely laugh this off, it was actually believed at one point that at least certain owl species could see in infrared, though this idea was put to rest in a scientific research paper that came out just one year before *All-American Comics* #25 (1939).[8] We can hardly hold it against his creators that Doctor Mid-Nite was based on a flawed understanding of owls that was apparently common at the time.

They also have the crafty Dr. McNider devise so-called "blackout bombs" to use as a weapon. These explosives work by producing a thick, black smoke that only Doctor Mid-Nite is able to see through. At first, this idea may seem a bit strange. How is the darkness resulting from a light source being *obscured* by dark smoke and soot – we sadly don't have the recipe for McNider's bombs – fundamentally the same as the *absence* of a visible light source? Well, fortunately for our inventor, infrared light is, in fact, more easily transmitted by smoke than visible light is, and if we accept that seeing in infrared is the key to his ability to see in the dark, this might actually work. Go figure.

There are notable differences between Matt Murdock and Charles McNider that are worth considering. Looking at how fictional blind characters are commonly conceived, the former is more typical of the "super-blind" than the latter. McNider has no heightened senses beyond his ability to see in the dark. Then again, his goggles appear to restore something close enough to natural vision that there is really no need to rely on the typical tropes.

Matt Murdock, on the other hand, has no prosthetic device he can use and Daredevil's core concept depends specifically on readers buying into the premise of compensatory heightened senses. And, despite the common assertion that Daredevil's blindness is merely a technicality, this isn't actually the case. Doctor Mid-Nite can in fact access the visual channel under special circumstances largely under his control whereas Daredevil cannot.

Projecting a darkness that works to the hero's advantage and blinds his enemies is a concept we also associate with another blind hero, this time from Daredevil's own fictional Marvel Universe: The Shroud. His darkness, however, is of a more mystical nature. Rather than being just a dark dust cloud, the Shroud uses a substance that hails from the ominous-sounding Darkforce dimension.

The Shroud, otherwise known as Maximillian Coleridge, was created by Steve Englehart and Herbe Trimpe, and first appeared in

1976, in *Super-Villain Team-Up* #5 (1975). The character's arc runs a full seven issues, but we don't actually learn about his origins until his third appearance. The Shroud spends most of the first two playing the part of the dark and mysterious stranger at the periphery of the main story which, among other things, includes then-Secretary of State Henry Kissinger signing a non-aggression agreement with Doctor Doom's Latveria!

In *Super-Villain Team-Up* #7, we finally get to the origin story which is told by the Shroud himself in flashbacks (his audience is the mildly interested Namor the Sub-Mariner who spends these panels resting in a bathtub). We learn that, as a young boy, the Shroud saw his parents gunned down before his eyes. This inspired him to devote his life to fighting crime, and as he grew up he studied criminology and science while also becoming an accomplished athlete.

Next, he traveled to Nepal and the ancient Cult of Kali where he was taught "the deadly arts, and the worship of their diabolical dark goddess!" Little did he know that the mark of success would be to receive the Kiss of Kali, which entailed being branded across the top of the face with a hot iron. After waking up in the snow outside the temple, where he had sought to ease the pain, the Shroud discovered that he was a changed man. He explains: "When I awoke, my eyesight was gone! That didn't bother me, because I no longer needed it – darkness was no different from light to me." If much of this story sounds like Batman's origin, it's not a coincidence. Steve Englehart himself has said that the Shroud was conceived as a mash-up of Batman and the Shadow.[9]

For our purposes here, though, it's interesting to note that very little is mentioned thus far about how and what The Shroud actually "sees," and his blindness is of little consequence to the story for the first few issues. In *Super-Villain Team Up* #8 it is even stated that he "... starts *backwards* – wide-eyed in *wonder* at something he *sees*!" which suggests that the creators had simply flat out forgotten the character was blind, and so soon after giving us his backstory! The Shroud also has his own airplane which he is somehow able to fly with no trouble at all (though it is regrettably destroyed by some Latverian peasants at the worst possible moment).

While the plot becomes increasingly crowded and convoluted, we finally get some insight into the Shroud's weaknesses by *Super-Villain Team-Up* #11 (1975), when the Shroud is dragged in front of a large monitor by the Red Skull: "No, damn it! I *can't* see! I'm *blind* — and my mystic senses *can't* tell me what's on that *screen!*"[10]

The Shroud's ability to summon darkness at will is not used at all in his first story, but is apparently added to the character's repertoire between his adventures in *Super-Villain Team-Up* and his next appearance in *Spider-Woman* #13 (1978), by Mark Gruenwald, Carmine Infantino and Al Gordon. When the Shroud formally introduces himself to Jessica Drew, in *Spider-Woman* #15, his origin story includes the mention of his eyesight being replaced by a mystical sense, and he also tells her that he discovered his ability to summon darkness following his last adventure. Interestingly, part of his reason for seeking her out, aside from their mutual connections to some previous plot points, is that he's unable to read due to his blindness and needs her help.

From a purely functional perspective, the Shroud's mystical sense can be understood as being similar to Daredevil's radar sense. It's not a form of sight that depends on light, but a way to gain a spatial awareness of one's immediate surroundings, and sense the shapes of nearby objects. The limits of this ability are explored in *West Coast Avengers* #1 (1984), by Roger Stern, Bob Hall and Brett Breeding, when Wonder Man, suspicious of the Shroud snooping around, attacks from the sky:

"The Shroud cannot see Wonder Man, for the Shroud is blind! His mystic senses only enable him to perceive objects that are closer than 100 feet. By the time Wonder Man is within that radius... it is almost too late to react!"

Daredevil's radar sense hasn't been quantified outside of *The Official Handbook of the Marvel Universe*, which gives an estimate close to that of the Shroud, but he has only rarely been shown sensing anything from great distances.[11] In this case too, the two characters appear quite similar. Where Daredevil has an edge is in the sensitivity and range of his other senses, as he would have more easily heard an attacker coming.

Like Doctor Mid-Nite, the Shroud is not quite the textbook example of a "super-blind" character that Daredevil is. While the Shroud shares Daredevil's predicament of not being able to see in a manner that comes close to traditional eyesight, he is also not endowed with any enhancements beyond his mystery ersatz "sight."

What Daredevil and the Shroud most definitely have in common, beyond their blindness, is that they are both fighters. They are physically adept, acrobatic and strong. This stands in stark contrast to some of Marvel's other blind, super-powered characters, who fall into a category best described as that of the blind "seer," another common blind literary archetype of which a well-known example is that of our previously mentioned Tiresias, the blind prophet of Greek mythology who has the ability to see the future, speak with the dead, and even understand birdsong.

Blindfold, also known as Ruth Aldine, is a blind mutant created by Joss Whedon and John Cassaday for *Astonishing X-Men* (2004). Born without eyes, Ruth is a telepath and clairvoyant. Cassandra Webb, the first Madame Web, is another mutant character with similar abilities. Webb, who is both blind and paralyzed, first appeared in *The Amazing Spider-Man* #210 (1963), by Denny O'Neil and John Romita, Jr. The abilities of Madame Web – along with her blindness – were later passed to Julia Carpenter, who started her Marvel "career" as the second Spider-Woman. The mutant character Irene Adler, also known as Destiny, is another example of a blind woman with powers of precognition. Similar characters appear in the DC comics universe, such as Blind Faith, an enemy of the Justice League who possesses telepathic powers.[12]

I find it interesting that most of the blind telepaths we know from comic books, including all those mentioned above, are women. Perhaps it is more difficult to think of male characters, at least those who have yet to reach middle age and beyond, as sufficiently powerful if they are not fully combat-ready. It may simply be the case that taking a physically more passive role is less harmful to a character's perceived femininity than similar weaknesses would be to the masculinity of street fighters like Daredevil and The Shroud. In *Sight Unseen*, author Georgina Kleege makes the following observation about the treatment of blind men in the history of cinema:

"Because blind men in the movies exist as passive objects for both the viewer and the viewer's on-screen surrogate, they perform the function that mainstream cinema usually reserves for women: They exist to be looked at. They are all spectacle. In treating blind men like women, movies reenact the castration that blindness has represented since Oedipus."

At first glance, this observation doesn't seem to apply to a character like Daredevil, who is very much the leading character of the comic. However, it does say something about the kinds of common prejudices *Daredevil's* creators were up against in introducing their new character to the masses. The readers quickly learn of Matt Murdock's abilities – his "not like other blind men" credentials – but the other characters in the comic do not. Matt's concerns about whether Karen's affection for him is real, as opposed to an expression of pity, and whether he can ever be good enough for her plague him throughout much of the early Silver Age era. *Daredevil* #8, in which Karen starts pressuring Matt to go to Lichtenbad for eye surgery, has Matt contemplating the outcome:

"It would mean the end of *Daredevil* as a force for justice! But it would be the only way I could dare try to make Karen my wife! For, I could never ask her to marry a sightless man!"

THE "IN-BETWEENER"

Attentive readers – especially those of you with an intimate knowledge of the greater Marvel Universe – may have noticed the absence of a certain character from the preceding list of blind heroes and villains. He is a short, homely sort of guy who surrounds himself with monsters – most of his own making – and has made a home for himself far underground.

I am, of course, thinking of the Mole Man, also known as Harvey Rupert Elder, the first official villain of the Fantastic Four! There are a couple of things that make Mole Man particularly noteworthy, and relevant to the topic of this chapter. One is that Mole Man is the first Marvel character to possess a "radar sense," and the only such character besides Daredevil. Mole Man made his debut in *Fantastic Four* #1, by Stan Lee and Jack Kirby, which hit the stands in the fall of 1961 and thus predates Daredevil's first appearance by over two years. With this timing of events, one can imagine that at least some of the thinking that went into the creation of Mole Man was repurposed for the creation of Daredevil.

According to his origin story, Harvey Rupert Elder was a man ostracized from his community on account of his hideous appearance. When he could stand this treatment no longer, he went in search of the center of the Earth. Yes, the literal center of the Earth. (As one does.) Finding, at last, a deep cavern on the aptly named Monster Isle, Elder fell to the bottom of the hole and, upon regaining consciousness, discovered that he had lost most of his sight. Being trapped underground, "like a human mole!" he took up the Mole Man moniker and "carved out an underground empire!" When he meets Reed Richards and Johnny Storm of the Fantastic Four, Mole Man describes his newfound abilities as follows:

"I conquered everything about me! I even learned to sense things in the dark – like a mole! Here, I'll *show* you! Try to strike me with that pole! *Try it,* I say!! Hah! I sensed that blow coming! Nothing can take me by surprise! And, I have developed *other* senses too like those of the bat — I possess a natural radar sense... A warning system which enables me to evade whatever danger strikes at me! Compared to the Mole-Man, you are slow... clumsy!! Hah hah!!"

In his next appearance in *Fantastic Four* #22, Mole Man once again puts his abilities on display in a battle against the Thing where he turns the lights off to gain an advantage: "See how easily I can plunge my domain into *darkness*! [...] You *forgot* that I can function in the dark due to my highly developed *radar sense!*" A note from the editor clarifies the situation for new readers at the bottom of the page: "Radar sense: Although his vision is weak, the Mole Man is able to sense things in the dark, as fully explained in F.F. #1."[13]

The way Mole Man's radar sense is described suggests that the term "radar" is either an outright misnomer – in that neither moles nor bats use actual *radar* to navigate – or used as a metaphor for a general ability to sense the spatial arrangement of objects without sight. In fact, it has long been a suspicion of mine that Daredevil's radar sense was not originally meant to be understood literally either. A recent discovery, which I will return to in chapter nine, has further convinced me of this. For now, suffice it to say that navigation by echolocation and man-made sonar both rely on the detection of reflected sound. Radar, which accomplishes the same thing using electromagnetic waves, has no biological equivalent.

The second thing that makes Mole Man's case particularly noteworthy is that he is *not* totally blind. This sets him apart from every other fictional character mentioned thus far. And, it is not a coincidence so much as part of a broader pattern. Fictional portrayals of total blindness are far more common than portrayals of people with less severe vision impairments. To again quote Bolt: "[T]he vast category into which most people with impaired vision fall is at best rendered temporary and at worst not represented at all." M. Leona Godin makes a similar observation in *There Plant Eyes: A Personal and Cultural History of Blindness*:

"Blindness and sight – as well as their analogs darkness and light – constitute, in the Western imagination, a fundamental dichotomy, but that has not been my experience and it is not the experience of most blind people, a very small percentage of whom were born with absolutely no sight."

The Mole Man character is, in effect, more of an outlier than he should be. In reality, the absence (or near-absence) of any light perception is far less common than milder forms of vision impairment.

Many people are familiar with the term legal blindness, which is used in the United States for tax purposes and to determine eligibility for certain services. To be considered legally blind, a person will either have a best-corrected visual acuity of less than 20/200 or a field of vision restricted to less than 20 degrees. The legally blind constitute around one-third of the total number of people with low vision in the United States, with low vision defined as having a best-corrected visual acuity of 20/40.[14] Among the legally blind, a sizable majority have at least some useful vision. Relatively few share Matt Murdock's condition of having no light perception at all.[15]

In this sense, most depictions of blindness in fiction are not only a poor representation of the lives of actual blind people but collectively serve to give the general public a very skewed sense of the phenomenon of vision impairment more broadly. With this in mind, it is not surprising if the average person, with little personal experience of vision impairment, is prone to thinking of total blindness and normal vision as sharply defined binary categories. In this simplified view, one is either totally blind or fully sighted.

I would argue that this particular fallacy is likely one factor that contributes to making Daredevil's sensory world, with its patchwork of enhancements and deficits, so challenging for readers and creators alike to fully wrap their heads around. Blindness is also something we tend to associate with certain practices and paraphernalia, such as reading braille and using a white cane or a guide dog. If we are to assume that Matt Murdock uses these only to maintain a secret identity, then readers can be forgiven for raising the question of whether he is blind at all. I will give the real and imagined implications of Daredevil's blindness a thorough examination near the end of the book.

It is clear that blindness is one of the more common, and quite possibly *the* most common, physical impairments among comic book superheroes and villains. The only possible exception would be the many characters who have seen amputated limbs replaced by futuristic prostheses – technological or biological – that always seem better than the real thing.

There is obviously something about the idea of blindness that inspires the imagination, especially when it is assumed to entail a complete lack of vision. Fictional characters as diverse as Oedipus, Zatoichi, and that incredibly creepy girl from Stephen King's *The Langoliers*, all have something in common with our own Matt Murdock in that they are manifestations, albeit diverse and largely erroneous, of long-held notions about blindness.

In other respects, however, Daredevil is also a fairly typical representative of a Silver Age superhero in the Marvel tradition. Some of the ways in which his body attempts to compensate for his blindness may have Wells' *The Country of the Blind* written all over it, but the means by which he achieved his powers came straight out of a barrel of radioactive waste. Or at least, that was how this pivotal event was described way back in 1964.

With the passing of time, radiation has lost its appeal as the transformative agent behind these kinds of miraculous transformations. Both of Daredevil's major live-action outings – the 2003 movie and the 2015-2018 television show – are awfully vague about what *exactly* caused young Matt's remaining senses to become dramatically heightened in the same accident that damaged his eyes. This makes sense, of course. Attempts to explain how all of this happened are bound to disappoint. No member of the audience would be made any wiser, and pretend science jargon always runs the risk of coming across as silly.

We should remind ourselves that any meaningful discussion of the science of modern superheroes and their powers requires that we move past this point of origin, that "miracle exception" I referenced in the book's introduction, and accept that certain things just are what they are. Any explanation of how young Matt Murdock would have been bestowed with dramatically heightened senses falls into the realm of the supernatural, just as no amount of sci-fi level genetics will really give the X-Men mutant Kitty Pryde the ability to phase through walls.

However, unlike many other similar heroes, Daredevil presents us with the opportunity to at least tackle the "what?" of the miracle exception, even when we cannot hope to make a dent in the "how?" Yes, having a reasoned argument about what changes would have to happen to young Matt Murdock's anatomy to get us at least part of the way to our destination is something that *can* be done.

First, we are going to look at what separates the abilities that follow naturally from the idea of heightened senses, and the ones that do not. We have to separate the sense from the nonsense.

CHAPTER 2
SENSE, NONSENSE, AND THE STIMULUS

 "All we have to believe is our senses: the tools we use to perceive the world, our sight, our touch, our memory. If they lie to us, then nothing can be trusted."

American Gods, by Neil Gaiman

When I was in the eighth grade, my social sciences class went on a trip to a local Hare Krishna center. Despite being a precocious fourteen, I remember fearing brainwashing as if it were advanced hypnotism. I dreaded the visit, thinking that it would put me under some kind of spell. As I got older, I realized that it usually takes a little more than a one-hour group conversation with the members of a religious sect to become a convert. This is especially true when the group in question is in the habit of getting up at four every morning (that would be a deal-breaker right there).

However, the sense of danger I felt at the time made sure that I can still recall this visit in great detail. One thing that has stayed in my memory was when our guide told us about a daily ritual that consisted of feeding the resident Krishna statue. To our group of skeptical middle schoolers, this "feeding" sounded more like "presenting food to," and one of my classmates raised his hand to ask whether the believers were of the opinion that Krishna actually sampled the food personally. Our guide responded enthusiastically that he was sure of it: "I believe Krishna can taste the food with his eyes; see it with his ears!" As you might expect, his young audience remained unconvinced.

The reason I bring up this seemingly random anecdote in a book about Daredevil's senses is not to launch into a discussion about what sorts of miraculous phenomena might exist beyond what is open to exploration by scientific means. Rather, I've chosen it because it provides an interesting perspective from which to view the senses in general. How so? Well, one question we might ask ourselves – as I did at fourteen – is whether it even makes sense to think of taste as an experience that can be had at all without a physical mouth to do the tasting. A mouth is, after all, equipped with a tongue covered in taste buds, and located downstairs from a nose that can detect the molecules wafting up through the back of the throat. Exactly the kinds of things you would think essential to experiencing the flavor of foods. If this sensation can be removed from such biological constraints as the mouth itself, what then is it really to taste something? Or to see, hear, and touch, for that matter?

Our sensory experiences are inextricably linked to our bodies as we move through and exist in the world, and come into contact with various kinds of energy and matter. But that's not always how we talk about them. In everyday conversation, the word "sense" gets to perform all kinds of duties.

According to the Oxford Dictionary, the first meaning of the noun form of the word is: "A faculty by which the body perceives an external stimulus; one of the faculties of sight, smell, hearing, taste, and touch." The second meaning is the much flimsier: "A feeling that something is the case."[1] Used this way, the word "sense" can convey anything from a well-educated guess to the vaguest of hunches. I would argue that the semantic gap between these two meanings of the word helps to open it up to something vaguely magical that probably appeals to writers of fantasy and science fiction, balancing as it does between the strictly physical and the metaphorical.

For instance, what exactly is it Darth Vader, of *Star Wars* fame, experiences when he "senses a disturbance in the force"? We're invited to believe that he is detecting something that belongs in the physical realm – I mean, it's a *"force"*! – which is in line with the first interpretation of the word. However, by trusting the audience to be just as familiar with its second meaning, a feeling that something is the case, a storyteller may get away with not having to explain the nature of that feeling.

We can ask the same question of Peter Parker's spider sense which famously warns him of impending danger. While we're all familiar with the bad "vibes" that come with places and people that put us on high alert in real life, this is obviously not due to some special sixth sense. The spider-sense, on the other hand, must certainly be something different and more powerful. But what exactly? It's described as a form of precognition, but is still – at least in the common understanding of the phenomenon – meant to be a physical sense that causes a tingling sensation. It is literally the second dictionary definition of the word 'sense' masquerading as the first. This is not an indictment of Spider-Man as a character. The superhero genre is replete with characters whose powers are much more esoteric than Peter Parker's. This is merely an observation that our powers of perception are often described in ways that allow a subtle conflation of the tangible and the metaphorical.

Daredevil's heightened senses are ostensibly intended to be of the more down-to-Earth variety in that we can expect them to adhere to a stricter understanding of the word "sense." The desire on the part of most writers and artists to attempt to *explain* how it is that Matt is able to do what he does – rather than deferring to some kind of Shroud-like dimension-bending or extra-sensory perception – is also evident in the Daredevil comic. This is particularly true for the first two dozen or so issues, where our hero seems to be informing the reader, through internal monologue, of his every sensory experience.[2] (And yes, it gets old pretty quickly.)

That is not to say that Daredevil's senses have never strayed from this path. Setting aside the many times they've been stretched to implausible degrees, let us not forget the time Daredevil sensed The Owl's "aura of unmistakable villainy" (*Daredevil* #3), as well as his "birdlike emanations" (*Daredevil* #22), to take just a couple of examples. Still, the case remains that most descriptions of Daredevil's power set are relatively straightforward. He was in an accident that blinded him and heightened his remaining senses.

I'm going to briefly stress the term "remaining senses" here since this actually draws a line in the sand for what we can and cannot expect this character to be able to do. His senses are *heightened*, yes, but those heightened abilities depend on his physically real and otherwise typical human senses. This means that we can assume that he senses the same kinds of things that we all do, with the obvious exception of light, which he doesn't sense at all.

Even the enigmatic "radar sense," to which we will devote most of the final part of the book, is commonly understood as being physical in nature, to the extent that it is viewed as a separate sense at all. As we will see in chapter nine, a strong argument can be made for the radar sense being inspired by real-life phenomena, and that the term "radar" itself was meant as a metaphor.

With all this in mind, we can safely conclude that whatever lapses in logic or errors in execution *Daredevil* creators have served up over the years, the foundational idea of Daredevil's power set is that his senses are of a physical nature, and similar to our own in every way but the degree to which he is able to extract information via the sensory channels – or sensory modalities – that he shares with most other humans.

The differences between Daredevil and a person equipped with the typical set of senses are quantitative rather than qualitative; not primarily a matter of apples and oranges, but one of apples and *more* apples.

THE "FIVE" SENSES

One of the "facts" we all learned as kids is that there are five human senses. We even see them listed in the first Oxford dictionary definition of 'sense.' This idea of the five senses is also reflected in the *Daredevil* comic. When we first encounter the character in *Daredevil* #1 (1964), young Matt tells us that:

- "My *hearing* is so acute, that I can tell if someone is in a room with me just by hearing the *heartbeat*!"
- "And I never forget an odor once I *smell* it! I could recognize any girl by her perfume... or any man by his hair tonic..."
- "Even my *fingers* have become incredibly sensitive! I can tell how many bullets are in a gun just by the weight of the barrel..."
- "While my sense of *taste* has become so highly developed that I can tell exactly how many grains of salt are on a piece of pretzel..."
- "But my most *important* new ability is in the form of a built-in *radar* that I seem to have developed! It enables me to walk anywhere safely, without bumping into anything!"

With the radar sense as a convenient stand-in for vision, this list fully recapitulates the idea, which can be traced back to the Greek philosopher Aristotle, that the fundamental senses are hearing, smell, touch, taste, and sight.

However, even a closer look at the abilities Matt lists above gives us a hint that the actual state of affairs is considerably more complex. For instance, when Matt explains that he can count the number of bullets in a gun, this probably strikes most of us as a rather odd example. First of all, you would think that the weight of an individual bullet in the barrel of a gun would depend on the make and model of the gun in question. Secondly, even if Matt did know all he needed to know about the gun, what he is doing is better described as *weighing* it in his hand.

Grasping and holding the gun is a tactile task, but it's not a test of the touch sensitivity of the hand and fingertips *per se*. In fact, the ability to precisely gauge the weight of an object has relatively little to do with the more superficial touch receptors in the skin and more to do with those internal stretch sensors attached to the muscles and tendons. These also help us keep track of our limbs in space, and in so doing they feed into something called proprioception, or kinesthesia, arguably a "sense" in its own right.

On the flip side, when we look at Matt's ability to count the number of grains of salt on a piece of pretzel, we may be wondering whether he is actually *tasting* the amount of salt or using his tongue to *touch* the individual grains and counting them? Most likely, it would be a combination of both, especially since grains of salt tend to vary in size. Since this particular ability is never used again outside of this context, we can never really know. (There is a reason this book doesn't contain a chapter devoted specifically to the sense of taste.)

This is further complicated by the fact that what we call "taste" isn't that easy to pin down. In the strict physiological sense, we use the term to refer only to what the taste buds on the tongue and inside of the mouth are doing when detecting something salty, bitter, acidic, sweet, or savory.[3] This isn't how most people use the word in everyday conversation, where it might instead refer loosely to the more complex experience of *flavor*, which Merriam-Webster defines as "the blend of taste and smell sensations evoked by a substance in the mouth."

Our full experience of food is further enhanced by its texture, or even the sound of it when chewed. If I wanted to be charitable to the aforementioned Krishna statue, I would even admit that so does the visual appearance of the food. In fact, even wine specialists have been fooled by the addition of red food coloring to white wine.[4]

Touch isn't one singular thing either. It's easy to think of it as any sensation that is felt by the physical body, but the special receptors that respond to light touch are not the same as those that respond to pressure, and they also vary in density and distribution across the body. Pain is another sensation that is very much its own animal, as is the sensation of hot and cold. Counted this way, there are clearly more senses than five. On the other hand, you could slice things differently. We might, for instance, categorize the full set of human senses by sensory receptor class, that is by which kind of "thing" can be detected, and end up with another kind of list:

Chemoreceptors, associated with smell and taste, interact *chemically* with specific molecules. Chemoreceptors are also associated with the important internal sensors in our bodies that gauge, for instance, the level of oxygen and carbon dioxide in the blood. *Mechanoreceptors* react to physical deformation, such as bending, stretching or pushing. As such, they form the basis of hearing, balance, and the various forms of touch. Proprioception, the aforementioned sense of our bodies and limbs in space and in relation to each other, also relies on mechanoreceptors found in muscles and tendons. *Thermoreceptors* detect the flow of hot and cold in the skin and certain other places in the body, and *photoreceptors* detect – you guessed it – photons of light to give us vision.

Perhaps the most basic property that the senses have in common, no matter how we choose to categorize them, is that they are all manifestations of some facet of the body's interactions with energy and matter. They detect very different things, but they all detect *something* – a stimulus. While this may seem obvious, it is at the heart of what sets us, and presumably Matt Murdock, apart from the many characters in fiction that sense things – the future, or other people's thoughts – by means that we cannot even begin to explain.

How and what we sense, and the type of information we can extract in the process, naturally depends on the interaction between the senses and their respective stimuli. At its most basic level, this information is constrained by the nature of the stimulus itself in ways that *cannot* be overcome by having sensory superpowers of the kind described for Daredevil.

UNDERSTANDING THE STIMULUS

Key to understanding *how* we sense is to first have a firm grasp of *what* we sense. This is particularly important to the understanding of Daredevil, since it entails the unusual exercise (for most of us) of having to disentangle the senses from each other and reflect on the significance of removing one of them. Many of the experiences we think of as being closely associated with a particular sensory modality are actually "multi-modal" to various degrees, as in our earlier example of flavor.

Light is by definition that portion of the electromagnetic spectrum that most of us, but not Matt Murdock, are able to see. It carries information about light sources, as well as about the surfaces that reflect light, and *how* those surfaces reflect that light. Light is able to convey very detailed information about the boundaries of objects and their surfaces because its wavelength is so very small in comparison to the size of these objects, ranging from 380 to 750 nanometers (nm), or less than one-thousandth of one millimeter. Because light travels fast (at nearly 300 *million* meters per second), it is also able to rapidly convey information about movement. Light can also travel unimpeded through empty space, which is how we can see stars on the other side of the galaxy.

Sound, meanwhile, carries information about particular kinds of events that disturb the air or some other medium. As in the case of (visible) light, what counts as "sound" in the general use of the term is limited to the particular range of vibrations that humans typically hear. Among the type of information we can extract by having access to the world of sound are the intensity and duration of the "disturbance," as well as other features of sound that separate one kind of disturbance from another, such as the constituent frequency pattern, and how this pattern changes over time. This way, sounds can tell us a great deal about the nature of the events that cause them: Is what we are hearing the honking of a car horn, or did someone drop a plate in the kitchen?

Like light, sound can also interact with objects and their surfaces in revealing ways. An object may either transmit incoming sound (a fancy way of saying that it lets the sound pass through it), absorb it as energy, or reflect it. In real-world situations, it may be some combination of the three. Just like surfaces differentially absorb and reflect light of different wavelengths, objects behave differently depending on the frequency of the incoming sound. As a general rule, low-frequency "bass" sounds are transmitted more easily than high-frequency sounds, which is one of the reasons it is so hard to shield yourself from the sound of jackhammers in the street. The lower the frequency of the sound, the less of it is also absorbed by the air or other medium, which allows it to travel farther.

There are, of course, important differences between sound and light. While both are wave phenomena, they are otherwise quite different beasts. Electromagnetic radiation, such as light, moves in the pattern of a *transverse* wave, like a sine wave being plotted along an axis.[5] In the case of light, the thing that is moving is the light itself, which can be thought of as a unit of energy, or a photon. (I'm probably breaking some kind of rule by calling the photon a "thing," since it has no mass at rest, and exhibits something called wave-particle duality which means that it's both of those things at the same time. We don't need to think too hard about this.)

Sound, on the other hand, does not fundamentally "exist" if there is no medium – such as air, water, or some other material – that can be set in motion by the event that gives rise to it. The medium acts similarly to a spring and moves in a *longitudinal* wave pattern that passes the sound along. Imagine that instead of looking at an up-and-down, alternating sine wave along the x-axis of a simple graph, you have a banded pattern that represents the atoms or molecules of the medium being maximally squished together, interspersed with areas where they are maximally far apart.

The different wave patterns were illustrated in memorable fashion by one of my physics professors in college using his own body as a tool. To show the movements of a transverse wave, he would make small steps forward while moving his hips in alternating directions, as if attempting to dance the cha-cha-cha. To convey the movements of a longitudinal wave he would instead move forward in small, shuffle-like jumps. As demonstrations go, it was pretty spot on!

Another practical difference between sound and light is that audible sound waves have much longer wavelengths than waves of visible light, which makes them considerably less able to convey precise spatial information about objects compared to light waves. This follows from a general property of wave phenomena that makes it impossible to resolve features that are smaller in scale than the wavelength.

What, then is the wavelength of audible sound? You may have noticed that I described visible light in terms of wavelength, whereas sound tends to be described in terms of frequency, using the unit Hertz (Hz). This is really just by convention as the two properties are closely related and can easily be converted from one to the other as long as you know the speed at which the wave travels. The speed of sound in air at room temperature is about 343 meters per second, which means that the wavelength of sound at the lower limit of human hearing (20 Hz) is equal to 343 divided by 20, or just over 17 meters. At the upper limit of human hearing (20,000 Hz), the wavelength is consequently 0,017 meters or about two thirds of an inch.

Because sound moves slowly (relatively speaking), some interesting and quite useful peculiarities arise when this stimulus "interfaces" with our sense of hearing. For instance, a sound coming from the side will reach one ear a tiny fraction of a second before it reaches the ear on the opposite side of the head. The difference is small, but not so small that our brains cannot tell the difference, and this contributes to our ability to hear the location of sound sources. The sluggishness of sound also allows us to hear echoes – or sound reflections – as distinct from the source of the reflected sound.

On a more philosophical note, when thinking about how we perceive light and sound, respectively, it is worth noting that our visual world is dominated by the "second-hand" light *reflected* off of objects and living things. We obviously wouldn't be seeing any of them without sources of light such as the sun, or artificial lighting, but these light sources usually appear to us as just another object (or celestial body) and are usually far less numerous than the objects they illuminate for us. The light sources don't steal the proverbial spotlight from light-reflecting objects unless, of course, they happen to actually *be* a spotlight, right above us, blinding our eyes with their intensity.

The situation is reversed with sound, where the *sources* of sound reign supreme. Reflected sound is something most people rarely pay conscious attention to, unless it becomes unusually obvious, such as during a visit to a massive rock formation or a large cathedral. This doesn't mean that we don't passively attend to reverberations on the regular – we do! – but it is generally the case that sound sources outmatch echoes both physically, and perceptually. Echoes are typically relatively faint, and the smaller and more distant the reflecting object is, the fainter the echo. And, because the presence of echoes detracts from our ability to locate sounds, faint as they may be, our brains actually work to further suppress them to some extent (more on this in chapter six!).

Of course, *one* of the reasons light and sound differ in this way is one we've alluded to already: the greater propensity of sound, especially of the lower-frequency variety, to pass through and around solid objects. It is true that some materials transmit *light* as well, in part or almost completely, allowing us to see through windows, water, and the other things we describe as transparent or translucent. However, the vast majority of the things we see around us do *not* have these qualities, enabling them instead to either reflect back or (just as conspicuously) absorb most of the plentiful ambient light.

Meanwhile, we find that hearing our neighbors having a party is a perfectly normal consequence of the world of sound. But with more of that sound energy passing through the wall, relatively less will remain to reflect the sound back at the party-goers. How to think about sound, echoes, hearing, and how we might construct a world for Daredevil that actually works within these constraints, are topics I will have reason to return to.

. . .

There is one more type of stimulus to cover among those that can provide a particularly rich source of information about the external environment. I am talking about the chemicals that serve as the stimuli for taste and smell, and give us vital information about whether something we are about to eat is nutritious or poisonous, or warn us of other dangers. Scents also do many other things for us, such as single-handedly help us identify places, people, and objects. Using nothing more than your nose, you can easily tell the difference between the beach, and your local Starbucks. Your grandmother has a particular smell, which is (hopefully) very different from the smell of your dog. A rusted metal can and a rubber ducky will smell very different. You get the idea.

However, like light and sound, scent is not some infinitely powerful entity, whether we are Daredevil or not. Odorous molecules differ greatly in their "volatility," that is their propensity to take to the air so that we can smell them in the first place. Many materials have no scent at all. While most people are familiar with the concept of an odorless gas, it may be surprising to learn that solid metals are odorless as well. What we call the smell of "metal" is really the scent of other compounds that form when metals undergo certain chemical reactions, including with substances in the skin.

Regardless of the volatility of whatever chemical compound we happen to be smelling or tasting, it is always the case that these molecules move much more slowly than light, and even sound, which is one reason they are poor conveyors of motion. The speed at which scents travel depends on several factors, including the size of the particular odorous molecule (the bigger it is, the slower its rate of diffusion through the air), and a more concentrated odor will obviously permeate a closed space more rapidly than a less concentrated source. If there is a flow of air, that will also greatly affect how scents travel. All of this makes it difficult to pin down a fixed "speed of smell."[6]

Scents are also "spatial" in ways that are much more vague, or dare I say flimsy, than light and sound. Scents can have a quality of over-here-ness, or over-there-ness, or they can appear to permeate a space. We usually deduce the source of scents by moving our heads, or whole bodies, relative to the source. In her excellent book *Smellosophy*, cognitive scientist and philosopher A. S. Barwich points out that:

"The nose does not convey spatial features of perceptual objects like the eyes do. But the nose allows us to behave spatially in relation to these objects," and "Smell has a clear exteroceptive, outward-oriented dimension: odors afford movement toward or away from something. And so olfaction is a spatial sense if understood as signaling external targets of behavior and in relation to an organism. But its spatial dimension must be understood via the sensory system, not the stimulus."

Seeing, hearing and smelling are all examples of sensory experiences where the stimulus, or at least some aspect of it, reaches us across space, to tell us something about an object and event that is physically removed from us.

To make the distinction, we call the object and event "out there" the *distal* stimulus. It is, in other words, at a distance. The more immediate thing that makes it possible for us to sense the object or event is called the *causal*, or *proximal* stimulus. For instance, an object in the presence of a light source is visible to us because it reflects photons, light "particles" (the proximal stimulus), that *traverse the space* between this object and our eyes, and physically interact with the photoreceptors in the retina.

In the case of sound and hearing, the rubbing of tires against asphalt, as someone hits the brakes on their car outside our window, is an example of a distal stimulus. The proximal stimulus is the vibrations of air caused by this event that set our eardrums in motion. When we smell something wonderful from the kitchen, freshly baked bread might be the distal stimulus, and the scent that reaches our noses – in the form of a microscopic sample of the bread itself – is the proximal stimulus.

It is sometimes subtly suggested that Daredevil can "reach out" with his senses. For instance, in *Daredevil #349*, by J.M. DeMatteis and Cary Nord, a caption reads "He casts his hyper-senses out like a net across the city – sifting through scents and sounds for the smallest hint of his teacher." However, this is not a great description of how sensation works. If there is any reaching to be done, it is the stimulus that has to do it.

Paying active attention to objects, sounds, and scents that are physically removed from us is something we do all the time, but it remains impossible to sense anything at all if there is not also some aspect of these objects and events that reaches *us*, directly or indirectly, and interacts with our sensory receptors. Our senses are not balloons to be inflated or deflated at will, which is another thing that separates a real sensory experience from simple comic book magic. Or the feasibility of feeding a Krishna statue.

Of course, not every sensory experience can be said to arise from a distal stimulus that is separate from the proximal stimulus. If you examine a textured surface using your sense of touch, the act of touching itself is what stimulates the appropriate touch receptors, as they rub against the surface. Taste also depends heavily on the physical contact between the stimulus, such as a piece of food, and the taste buds, collapsing the division between distal and proximal. In the case of balance, the stimulus is simply relative motion and acceleration. Vision, hearing, and smell, on the other hand, are largely devoted to relaying information from those parts of the outside world that are physically removed from us, and are often thought of as our distance senses.

As already mentioned, vision and hearing in particular give us a good idea of not only what is "out there," but the spatial arrangement of objects and events as well. Of course, there are many situations where a more "proximal" sense, such as touch, can and will detect a distal stimulus. Consider the case of an intense bass sound, which can be felt as well as heard. Or, when someone slams a door shut with such force that the displacement of air can be felt by someone close by. Our sense of hot and cold can also pick up on distal events, such as the opening of a window on a cold day, or when we are in the presence of a concentrated source of heat, such as a fire. Complications and gray areas aside, it is helpful to keep the difference between proximal and distal stimuli in mind when thinking about what we can sense, and under which circumstances.

Some of the more recent interpretations of Daredevil's enigmatic radar sense have suggested that it might be understood as an ability that rests on the combined efforts of his heightened remaining senses. However, if we keep in mind that the senses vary considerably in what kind of information they can give us about objects that are literally out of reach, it should be clear that not all of the senses can be expected to carry this load equally if they contribute at all. After all, not even Matt Murdock can taste the shape of a desk. (Well, unless he walks over to it and traces its surfaces with his tongue. In which case he would technically be touching it. He would perhaps be *tasting* wood stain and polish.)

STIMULUS PROBLEMS

One thing I have realized over my many years as a naturally incredulous Daredevil fan is that most of the things the title character does that have me scratching my head (and muttering "that's *not* how it works...") have nothing to do with the plausibility of heightened senses *per se*. I love Daredevil's heightened senses, and his general concept as a character. This is the main reason I have written tens of thousands of words about the topic at *The Other Murdock Papers* website and a big part of the reason you are reading this book right now. Much of what Matt Murdock does, from hearing heartbeats to sensing the presence of silent objects in the environment, is perfectly amenable to something close to a scientific explanation.

Instead, what typically causes these moments of head-scratching is when Daredevil's creators fundamentally misunderstand a thing or two, not about sensation specifically, but about the very nature of the various potential stimuli at Matt Murdock's disposal. The stimulus relied upon to solve a particular problem may be acting "out of character," and sometimes it's unclear what the relevant stimulus is even supposed to be. In this section, I will introduce you to this concept by looking at scenes from the comic and beyond. I will do my best to explain why they fundamentally do not work, in that they don't follow naturally from the elegant central concept of heightened senses.

Let's start with my favorite piece of bad senses writing, the infamous second issue of *Daredevil*, by Stan Lee and Joe Orlando, in which our hero successfully lands a spaceship after being placed in it, while passed out, by the villain of the month Electro. After Daredevil regains consciousness onboard the spacecraft, we are treated to a series of panels describing how he is able to steer the ship:

"Imagine a blind man operating a space ship!! Not as impossible as it seems if that man can hear the levers move, *feel* the power needed, *sense* the direction of flight!!" Matt himself chimes in: "By hearing the slight movement of the astro-compass, I can gauge my direction perfectly! And by feeling the action of the radarscope I can pinpoint my landing! I'll bring the ship down in the middle of Central Park in New York, finding an open spot where I hear no human heartbeats."

As mentioned earlier, Stan Lee did an admirable job of trying to put into words what his new superhero was doing and how, at any given moment. But, ambition can only get you so far. Hearing the levers move, and sensing the direction of flight is one thing (though questionable in space), but through which mechanism is he "feeling" the power needed? Are we back to vague hunches á la Darth Vader? More crucially, how is he doing anything with what he is able to hear from the "astro-compass" and "radarscope"? It is not inconceivable that their movements are audible, but these movements are only meaningful if there are unit scales or other indicators to compare them against, and in this case, Matt would only be able to do so by visual means. There simply aren't sufficient stimuli, of the kind that would be useful to *him,* to make any of this remotely plausible.

Rather than spend too much time focusing on a single early issue of Daredevil that is a particularly egregious example of Silver Age madness, let's look at a more recent example from a different medium. In the first season of the *Daredevil* television show, there is a scene in the sixth episode, *"Condemned,"* that sees Matt trapped in an abandoned warehouse, on the phone with his new friend Claire, who is a nurse. He is trying to keep the injured mobster Vladimir alive long enough to press him for information and needs her help. When she asks him what's around him, he responds: "Half a box of nails, broken glass, wood, duct tape, old roadside emergency kit, a lot of plastic sheeting." Claire asks whether there are any flares in the kit, and Matt says there are two.

If this scene had included Matt moving around the room to examine some of these objects, preferably by touch, nothing about it would have seemed strange to me. Instead, he remains in place throughout this conversation, moving his head slightly, while all of these objects are subtly being precisely detected and identified from a considerable distance of at least ten feet. To quote myself from the review I wrote of this episode after the show's premiere:

> "To be clear, I easily buy the broken glass (he's probably stepped in some and the windows are not what they should be), the wood (wood-smelling rubble?), and the plastic sheeting (it probably has a distinctive smell, there's probably some flapping in the wind happening and it obviously reflects the sounds in the room differently than a solid wall or windows would). I will even buy the duct tape, which may also have a distinct smell. The half a box of nails though? How would he know that's nails as opposed to screws or washers or whatever else from clear across the room? And the two flares? Does one flare have one specific unit of scent or something?"

Contrary to what you might suspect from reading the above passages, I actually appreciated how Matt Murdock's senses were handled throughout much of the *Daredevil* television show. I was especially impressed by the show's characterization of Matt's blindness, which is something we will have reason to return to. However, the scene above stands out, along with a handful of others, in being virtually incomprehensible from a sensory perspective. Nothing that happens during this scene, or in any of the episodes leading up to it, explains to the viewer how Matt accomplishes this. The audio description of this part of the episode offers us the narration: "He lowers the phone and puts his senses to work." Certainly, but *how*? This explains nothing.

The problem here is not that Matt can sense the presence and shape of silent objects in general – he wouldn't be able to be Daredevil if he didn't – but that the decision to have this scene unfold as it did appears to lack any understanding of what kinds of stimuli might serve as the *basis* of such an ability. We have a collection of distal stimuli, but no sense of what the proximal stimuli could possibly be. Walking around the room, and approaching the objects, would not only make sounds that can differentially be reflected by the objects and back to Matt's ears, but tell him more about the exact source of whatever scents he might be smelling.

Interestingly, in the second season of *Daredevil,* there is a scene in episode eight, *"Guilty as Sin,"* in which Daredevil and Elektra are confronted by a gang of ninjas, that completely contradicts this scene from season one, along with most others throughout the show's run. To his dismay, Matt realizes that he is unable to sense the ninjas' presence and location since they are able to hold their breaths and quiet their heartbeats. This would suggest that objects need to be active sources of sound for him to be able to detect them. The problem is that this notion completely undercuts Matt's ability to be Daredevil in the first place, which obviously presupposes that he can detect otherwise silent objects in the environment.

When you put scenes like these side by side, it presents us with an excellent example of why there is so much to be gained from letting real science help guide the way Daredevil is written. I fully realize that the science itself is not as important to every fan as it is to me, but most fans do care about consistency. Thinking long and hard about what works, and why, saves writers from introducing unnecessary inconsistencies into the storytelling. Instead of having a situation where a character's abilities vary wildly from scene to scene, some foundational principles set us up for a more robust understanding of the character that is less confusing for viewers (or readers of the comic). The only reason I can think of why creators aren't already doing this is that they don't realize it can be done.

The beauty of Daredevil is that while you definitely need to buy into the idea of heightened senses to account for his abilities, you don't have to treat them like magic. In fact, doing so completely undercuts the simple elegance of the underlying concept. Daredevil is an eminently explainable character who is particularly ill-served by hand-waving and "whatevers."

The first two examples of "stimulus problems" I've mentioned above highlight cases where Daredevil is able to solve sensory problems in ways that are fundamentally unclear, or seem largely nonsensical. A more narrow kind of stimulus problem arises when we know which sense is *allegedly* being used, but where the stimulus simply isn't up to the task. There are many examples of this, affecting all the senses in various ways, but for now, I will restrict myself to smell, and let that serve as an example of a broader point.

In the case of smell, we are dealing with a sense that went from being criminally underused for at least the first fifteen years of the *Daredevil* comic, to becoming slightly less so over time. Still, when this sense is not being forgotten or undersold, it has occasionally seen some pretty extreme applications that don't line up with what we know about how the scent stimulus behaves.

As mentioned earlier, smell is generally considered to be a distance sense. Scents can and do travel from their respective sources to our noses, though it takes longer for an odorous molecule to reach us than for sound or light to do the same. Unlike sound and light, which travel "through" the air, odors spread *with* the air. And, because they do so, odors cannot readily move through barriers that air or other mediums cannot.[7]

One example of a scene that appears to contradict this appears in the second episode of the first season of the *Daredevil* television show. After having been tended to by his new nurse friend Claire Temple, Matt suddenly notices a person "on the third floor," going door to door asking questions, and comments on his accompanying smell of cologne.

The problem here is not that Matt can hear this person. If you ask me (and perhaps even Matt himself), I would say that this, rather than the cologne, is what he is actually responding to. Scents, on the other hand, can only reach the inside of the apartment where there is a flow of air. Of course, walls and other materials can have small cracks or be sufficiently porous to allow air and odorous molecules to pass, but this takes a very long time and is naturally highly limited in scope in a situation like this one.

Despite Claire's later assertion that her new mysterious acquaintance can "smell cologne through walls," that cannot actually be what he is doing. Matt would be able to smell it through the gaps between the door and the wall, though, or perhaps the ventilation, and that is where the scent will appear to be coming from. If he identifies the smell of cologne as coming specifically from the third floor, it is only because he associates this novel scent with the sounds of the stranger wearing it. The problem isn't Matt's nose, it is the nature of the stimulus, and how it interacts with the medium that "carries" it.

As examples go, this is not a particularly egregious one, but it is illustrative of a larger point. For reasons mentioned earlier, we cannot even under the best of circumstances, and with nothing blocking the flow of air, immediately detect the source of a scent with any great degree of precision. The context will help us associate the smell of smoke with the sight of a fire or the smell of coffee with the Starbucks we are approaching on foot. However, even bloodhounds *track* scents, they don't just run straight to their targets.

With this in mind, we can take a look at an example that is far more extreme than the relatively minor confusion we saw in Claire's apartment. In *Daredevil* #10 (2015), by Charles Soule and Ron Garney, Matt comes face to face with a morbid piece of art created by the serial killer Muse. The piece is a vast painting made with the blood of Muse's victims. While Matt obviously cannot see the painting, he notes with some confidence that the blood came from "at least one hundred and thirteen" different victims!

There are two major problems with this scene. Which one is the bigger one depends on what assumptions you make about what has happened here. In the first scenario, let us assume that Muse has pooled all of the blood from his different victims into one bucket so that they are all blended together before being used for his painting. How would you then identify the constituent parts (except with a DNA test), the way Matt apparently does in this scene?

You might assume that since we are all unique there must be some secret ingredient in each of our bloodstreams that might function as a signal of our unique biochemistry. We all have a unique scent, right? In this scene, Matt mentions "body chemistry, diet, background… all sorts of factors." While this is true, what is unique about each person's scent isn't typically any single "ingredient." What is unique is the *pattern* of the constituent scents. These constituent scents derive mostly from a common pool of biochemical compounds that are not infinitely variable on a molecular level. To the extent that they differ, they would do so unpredictably.

What Soule is asking of Daredevil in this situation is similar to any one of us performing a taste test of a tomato sauce mixed from a dozen different brands of store-bought tomato sauce (sorry about the visual!). Each individual tomato sauce may be unique, even when its constituent parts are not. They tend to all contain tomatoes, vinegar, basil, onions, and another handful of ingredients, including maybe a couple of different kinds of preservatives. Maybe one of them is super heavy on the garlic, another one has added chili, and a third has its blend spiked with thyme or marjoram. How would you possibly know how many different brands of tomato sauce went into that one big vat, especially if you've only ever tasted two or three of them? It may be that the brand with the chili in it is your favorite and that you are able to correctly identify this brand as definitely being in the blend. Good for you, but no cigar.

What I'm getting at is that this is at its core a mathematical problem. You are trying to solve an equation with too many unknowns in it. It can't be done. And, if any of this reminds you of the scene in the *Daredevil* television show where Stick and young Matt eat ice cream together and Stick lists how many dairies the milk came from, you are spot on.[8] The faulty logic is the same.

For a slightly different take on the scenario with the blood painting scene, we are going with the assumption that all of the sources of blood were *not* mixed beforehand, but applied one at a time so that a couple of brush strokes in one corner came from one victim, and all the other ones were similarly spatially separated. This makes the math problem a little bit easier since sniffing one small patch of the painting would be different from sniffing another patch some distance away. It makes it possible, in *theory*, to separate the blends from each other, even though it would be a painstaking (and morbid) process. However, while Matt does go up to the painting and presumably gives it a sniff here and there, he is by no means doing any kind of grid search.

If scents behaved like light, *and* Matt's nose had the image-forming properties of an eye, "scent-viewing" the whole painting could work in theory. (Of course, even those of us *with* sight probably wouldn't be able to immediately look at a complex painting and deduce exactly how many different hues of paint contributed to its completion.) However, as should be clear at this point, scent *isn't* light and shouldn't be expected to behave as if it were. At even short distances from the painting, the spatial separation of the different brush strokes would be so blurred that they may as well have been painted from that one blended bucket of blood!

We do not smell "scent-images." Again, this has as much to do with the *nature* of scents as it does any perceived inadequacy of our sense of smell. As mentioned earlier, Daredevil's sense of smell still tends to be underused, and I can think of a number of truly cool things you could use this sense for. None of them need include treating scent as if it were something else entirely.

For example, in *"Dogs to a Gunfight,"* the second episode of season two of *Daredevil*, Matt does some investigative work by using his nose. After visiting the bloody crime scene left behind by Frank "The Punisher" Castle in the previous episode, he catches the scent of a dog on a loose chain on the ground outside, and follows the scent of the dog's blood, in a stepwise fashion, to Frank's hideout. Once he gets nearer, he hears the faint sound of police radios. This is a vastly more credible sequence of events than any of the scenes I mentioned above. And this makes it more instinctively relatable.

The concept of the "stimulus problem" is one we will return to, but I hope it will be clear what it is I mean when I point out that a particular stimulus is misbehaving.

Of course, a large part of understanding how it all fits together is to look at how these stimuli interact with our bodies because while the existence of a stimulus is certainly *necessary* for sensation and perception, it is not *sufficient*. The task of forming the crucial link between the various stimuli in our environment on the one hand, and the central nervous system on the other, falls to a range of specialized organs and cells in our bodies. They dictate which kinds of interactions we are able to have with the matter and energy around us, with no exceptions – not even for Daredevil. It's time for us to turn from physics to biochemistry.

CHAPTER 3
BODY, MEET WORLD

" "There are many ways for organisms to probe the external world. Some smell it, others listen to it, many see it. Each species therefore lives in its own unique sensory world of which other species may be partially or totally unaware."

Richard Axel, from his 2004 Nobel Lecture

We should probably begin by taking the time to define the term "sensory receptor." I have used it a few times already, and you may even be familiar with it from high school biology. The terminology isn't entirely straightforward though. When we talk about a sensory receptor, we typically mean a specialized cell that may or may not be a nerve cell.

Nerve cells, or *neurons*, have some unusual properties that set them apart from most cells in the body, one of them being extremes of shape and reach. For instance, a *single* neuron responsible for receiving and sending touch signals can stretch from your fingertip, run up your spinal cord and make direct contact with other neurons in the brain! For this reason, you may occasionally see "sensory receptor" refer specifically to the bushy nerve endings near the skin of the particular nerve cell which typically detect the stimulus.

This part of the nerve cell consists of either "free" nerve endings – just the bushy little nerve fibers themselves – or encapsulated nerve endings. When they are encapsulated, the nerve endings work in a kind of partnership with specialized tissues consisting of other kinds of cells that aid in sensation. Most of the sensory receptors involved in specialized forms of touch, which can range from deep pressure to gentle vibrations, have encapsulated nerve endings. Our thermoreceptors, on the other hand, have free nerve endings that happen to be functionally specialized for detecting heat.

Olfaction, that is the sense of smell, is also conveyed by the free nerve endings of so-called olfactory receptor neurons. One really cool thing about these particular nerve endings is that they are the only ones in the entire nervous system that make direct contact with the outside world, protruding as they do through a porous piece of bone at the top of the nose to dangle freely. Well, there's technically some mucus to protect them and for the odor molecules to swim around in, but that's about it.

Specialized receptor cells that are *not* neurons are simply other kinds of cells that typically arise from early epithelial (skin) cells during development. While they are not neurons themselves, they form close partnerships with neurons that can receive the signals they send. The cones and rods in the retina are examples of specialized receptor cells, as are the hair cells in the inner ear, and the gustatory cells in the taste buds. This description of sensory receptors concerns their *structure*. It tells us what kind of tissue or cell type they are, basically. However, we have also previously talked about how sensory receptors can be classified by their function – what kind of "thing" they respond to – and concluded that the most relevant to human experience are chemoreceptors, mechanoreceptors, thermoreceptors, and photoreceptors.[1]

When we classify receptors functionally, we can get a bit of overlap in terms of *structure*: mechanoreceptors come in the form of neurons with special nerve endings (touch), and in the form of specialized receptor cells (hearing), for example. Chemoreceptors can also be neurons (smell) and specialized receptor cells (taste). But there's no need to worry! To figure out what ultimately makes receptor cells, or specialized nerve endings, able to detect any kind of stimulus at all, and to specialize for *particular kinds of stimuli,* we are going to zoom in so far that it doesn't matter what type of cell we're looking at. Forget picturing a wispy nerve ending or a rod-shaped... *rod.* We are going to get close enough to look at the cell membrane.

The cell membrane, as you may also recall from high school biology, consists of a "lipid bilayer," that is a double layer consisting of molecules whose "heads'" are water-soluble and whose dangly "legs" are fat-soluble, so that the molecules line up with their fat-soluble parts facing each other (playing footsie) and the water-soluble parts facing the inside and outside of the cell, respectively. The visual of two rows of hair pins might also be helpful here.

But this is not all that cell membranes consist of. Depending on where you look, you will find lots of specialized proteins that call the cell membrane home. Some of them are anchored to the inside of the cell membrane, some are anchored to the outside, and others still span the membrane so that one part is facing the outside of the cell and the other faces the inside. As you might imagine, cells can do all kinds of fun stuff with these proteins, including receiving messages from other cells or, crucially for our purposes, the outside world.

The proteins that specialize in receiving messages of various kinds are called receptor proteins. (Yes, I have just used the word "receptor" *again,* and yes, I know it's confusing. Just keep in mind that *sensory receptors* are cells that accomplish their various forms of "receiving" because they are equipped with special receptor *proteins.* Let's move on.)

In humans and other animals, the receptor proteins used by both photoreceptors (rods and cones) and olfactory receptor neurons belong to a "superfamily" of related proteins called G protein-coupled receptors (or GPCRs for short). As expected, these proteins sit in the cell membrane of the cells where they are expressed, with part of the protein facing the outside of the cell, and the other part facing the inside.[2] The part of the GPCR that is facing the outside has a binding site that can be activated by particular molecules in a lock and key fashion.

When the "right" molecule, in this context known as a *ligand*, comes along and binds to the GPCR, the latter changes its shape. This, in turn, will affect the behavior of its "buddy" protein (not a real scientific term) on the inside of the cell. This buddy protein (yes, I'm sticking with it) is a representative of another family of proteins – the G proteins. It, in turn, goes through its own changes upon receiving this important message from the outside of the cell and passes it on down the line through various chemical reactions.

Because G protein-coupled receptors are so diverse, a vast array of chemical messages can be relayed this way. The human genome contains approximately 800 genes that code for different GPCRs, and about half of these are expressed by olfactory receptor neurons, and are thus involved in smell (and a topic we will get back to in chapter seven). The remaining half of these genes code for GPCRs that are involved in other cellular processes throughout the body. The central nervous system uses receptors in this family to bind and respond to certain types of neurotransmitters, as does the immune system for much of its signaling. Of all the pharmaceutical drugs approved by the U.S. Food and Drug Administration, more than a third act on a GPCR. (If you are only going to learn the name of a single family of receptor proteins, for use at cocktail parties, this would be the one.)

It is easy to see how this family of receptors would form the basis of our sense of smell when the stimulus consists of distinct molecules that enter our mouths and noses and can function as ligands. But what about vision? Photons are ephemeral little packets of energy that can't even decide whether they are waves or particles. They don't *bind* to much of anything.

The way nature has solved this conundrum is quite remarkable. By pairing a GPCR in the "opsin" subfamily with a molecule called retinal, photoreceptor proteins can bypass the dilemma of not being able to bind photons directly by essentially binding to a light switch instead. Retinal, also known as retinaldehyde, is an organic molecule made from Vitamin A that changes its shape when it absorbs a photon.[3] This, in turn, changes the shape of the receptor protein and nudges its G protein buddy *transducin* into action inside the cell. What eventually ends up happening inside the receptor cells in both vision and olfaction, is a very abrupt change in the flow of ions across the cell membrane. You will recall that ions are molecules or, as in this case, atoms that carry a net charge. The most important ions involved in this flow across the cell membrane are potassium (K^+), sodium (Na^+), and calcium (Ca^{2+}) ions. This event functions as an electrochemical "trigger."

All cells, whether they are involved in sensation or not, contain tiny molecular pumps, which also sit in the cell membrane, that work to selectively pump certain ions into the cell, and other ions out of the cell. This causes a difference in the concentration of these ions to build up across the cell membrane. The point of having different concentrations of ions inside and outside of the cell is that it creates a difference in voltage and allows cells to function as microscopic batteries.

All cell membranes also contain ion channels, another kind of membrane protein, that can be either closed or open. When they are open, ions can flow freely "down their gradient" which naturally reverses the hard work of the ion pumps. And it really is hard work; a substantial proportion of the cell's energy is devoted to pumping ions in directions they are not inclined to go. The signaling cascade that started with the binding of a scent molecule or "light-activated" retinal in our earlier example ultimately acts on particular ion channels in their respective cells.

The behavior of photoreceptor cells is unusual in that they are in a kind of "on" state when there is no light striking them, and turn "off" when light hits. Olfactory receptor neurons, on the other hand, behave in a more typical fashion and open ion channels in response to a stimulus. Either way, there is a change in the status quo of the cell, and that is what is significant and acts as the trigger.

. . .

Using G protein-coupled receptors to send messages by triggering ion channels to open and close obviously works well enough. But ion channels can also be used directly to register an external stimulus — after all, they are also protein complexes that sit in the cell membrane, with some portion sticking out. Most of the other kinds of receptor proteins used in sensation are, in fact, different types of specialized ion channels.

For instance, what makes thermosensors out of a particular class of free nerve endings is the fact that their cell membranes have *temperature-gated* ion channels. That they are temperature-gated means that they open or close in response to heating or cooling. One relatively famous such ion channel that occasionally makes an appearance in the popular press is called TRPV1.[4] This channel is activated at temperatures greater than 43 °C (109 °F). It can also be chemically activated by the molecule capsaicin that is found in chili peppers. Hence that famous burning sensation!

Other types of ion channels are sensitive to actual physical distortions and are said to be *stretch-gated*. Such ion channels are found in the mechanoreceptors involved in touch, hearing, and balance where mechanical stress causes bending or stretching of some specialized part of the ion channel receptor which causes it to open. At the time of this writing, relatively little is known about the finer details of the transduction of sound. What *is* known is that the individual hair-like protrusions, or "cilia," that make up a small bundle on the top of each hair cell are linked by microscopic tethers called tip-links. When the cilia bend in response to incoming sound this tugs on the tip-links which causes ion channels to open and signal the activation of the cell.

The fact that we have cells in our bodies that can use amazingly specialized proteins to translate the wide array of matter and energy that impinge on our bodies into electrical signals is insanely cool. It is not, however, the result of anything that defies explanation. By accounting for the mechanism by which sensation actually happens, we can reach an understanding of what shapes our bodies' sensory abilities at their most basic level. Matt Murdock's senses, too, would be subject to this same kind of basic logic.

SENSORY ORGANS

Of course, the story doesn't end at the level of the individual cell and its particular micro-machinery. From our diverse arsenal of sensory receptors, entire sense organs are built that make use of them in structured ways, which facilitates further specialization. One key to the great variety of sensations we experience is that each individual sensory receptor typically responds to only one very thin slice of the energy or matter that makes up what we think of as reality.

Any single photoreceptor responds only to light within a narrow band of particular wavelengths and, as importantly, only to light that falls on the particular patch of the retina where this photoreceptor sits and is able to respond. A touch receptor similarly responds preferentially to a particular kind of interaction, such as light touch, or steady pressure. And, it only responds to the stimulus if it occurs within the patch of skin from which the receptor is able to detect it, that is its *receptive field*. A particular hair cell in the inner ear responds only to distortions at or near its location. Where this distortion occurs is determined in turn by the frequency of the incoming sound in ways we'll get back to later in this book.

The term "receptive field" was originally coined to be used as I just did above, to describe an area of skin that a particular touch receptor can detect input from. The term later came to be used for vision as well, and its meaning has continued to expand to be applied not just to a two-dimensional "input area," but as a broader descriptor of all the ways in which a receptor, or neuron generally, is limited in its response to very specific stimuli.

The size of the receptive field matters a great deal to how much detail can be resolved by a particular sensory system. The reason we can see in such fine detail at the very center of the visual field is because the photoreceptors here, exclusively cones in this case, sit very close together, with each photoreceptor relaying its information to a single neuron in the optic nerve. Meanwhile, the neurons that receive input from the rods are hooked up to many rods each, around one hundred on average. So, while the density of cones and rods is roughly the same, an individual cone can pass along much more detailed information. The way the retina is "wired" means that cone input is like the latest HD video game graphics, while input from rods is more like an 8-bit Nintendo.

Our sense of touch is also not an equal opportunity relayer of information. There is a much higher density of touch receptors in the fingertips than in the skin on the back, and the former have much smaller receptive fields. Why is it that we don't have such a fine sense of touch everywhere? Well, there are trade-offs to consider. The brain is an expensive organ with a limited amount of real estate. The same goes for the rest of the nervous system. Imagine how thick your spinal cord would have to be if every part of your body was as richly innervated as your index fingers. And for what? Fine touch on your kneecaps is wasteful and makes the whole system less efficient.

On the topic of tradeoffs, we should also point out that there can be explicit benefits to sensory receptors having larger receptive fields. What such a receptor loses in precision, it gains in overall sensitivity since it will be sampling input from a larger area. A perfect example of this can be found in the visual system where the aforementioned distribution of rods and cones combine to create something that is greater than the sum of its parts.

While cones are responsible for high-resolution color vision, rods are agnostic as far as color is concerned. What the rods have going for them is that they are much more light-sensitive, which makes them crucial to our ability to see anything at all in low light conditions. In addition to this, we just learned that the neurons that receive input from rods are connected to lots of rods, which expands the size of the receptive fields of these neurons and makes the eye even more light-sensitive.

Taken together, this accounts for two phenomena you are probably well familiar with. The first is that the world appears to us in shades of gray, rather than full color, when we observe it in very dim lighting because the cones are not active. The second is that, if you want to look at something in near darkness, you may have to shift your gaze slightly, rather than try to look directly at the object in question, since the very center of the retina contains *only* cones. The fact that we have different kinds of photoreceptors that are optimized for different tasks makes the visual system much more useful overall. Just like other parts of our anatomy, the senses are under evolutionary pressure to strike the right balance between function and efficiency in ways that maximize the ability of the whole organism to thrive.

Another somewhat random, but oddly Daredevil-relevant thing to know about rods is that even though they are "color agnostic" they still have a "favorite" wavelength that they respond maximally to. This wavelength happens to coincide with what our color vision system would label a bluish-green. This means that they respond less well to certain other wavelengths, say the ones that we classify as red. This means that red objects appear to be relatively darker in low light conditions than blue or green ones. This is known as the Purkinje effect, and makes for an excellent reason why Daredevil should ideally wear red when prowling the streets at night!

While we are on the topic of the eye, let us also stop for a moment and consider how important the structure of the sensory organ itself can be to its overall function. If we took this amazing retina we have just showered with praise and removed it from the back of the eye, all connections intact, and carefully placed it on the cheek instead, what do you think would happen? Would you be able to see an image of anything? The answer is a resounding no.

So-called eyespots are collections of photosensitive cells that represent an important early stage in the evolution of the eye. They work a little something like our hypothetical cheek-retina (although the human retina is, of course, vastly more complex). The problem with having a flat patch of photoreceptors just sitting on the skin like that is that a beam of light activates *all* of them at the same time, regardless of where the light is coming from. This of course extends to every light source, whether direct or reflected. It's like opening a file in Photoshop and applying "average blur." What you end up with is no image at all, but a solid color that is an average of all the hues in the picture.

The next step between this and a fully functioning vertebrate eye is to make an indentation for the retina to sit in so that the photoreceptors are at least shaded from light falling from some angles. We are still dealing with a complete blur, but one that may at least be a little brighter where most of the light is falling. This makes it possible to detect the direction of light, and moving light, but what we are seeing is not an image, by any stretch of the imagination.

If we continue by making the indentation deeper and start closing it up in front, the eye in our thought experiment will start to resemble something called a pinhole eye. The nautilus, a species of mollusk, is an example of an animal with eyes like these that function similarly to a pinhole camera. The pinhole constrains the amount of light that falls on the retina, and also naturally "directs" light from different areas of the scene in front of it to fall on different areas of the retina. We can now legitimately start talking about the eye forming some kind of *image*, but it is still very blurry compared to what we and most other vertebrates see.

What we have, and the nautilus doesn't, is a lens that sits behind the pupil, and a rounded cornea that makes up the front of the eye over the iris and pupil. Both of these structures, the lens in particular, take the pinhole eye and turn it into an optical superstar by precisely directing the incoming light. The role of a well-behaved lens in casting a sharp image upon the retina should be clear to everyone who wears corrective lenses. It is by combining the image-forming capacity of the eye with the fine makeup of the retina that the human eye is able to extract such rich and detailed information from the light stimulus itself.

.   .   .

Since Daredevil is a blind superhero, we probably don't expect his creators to get the very act of seeing wrong. And yet, this sometimes happens in surprising ways. One fundamental misunderstanding of the science that bears mentioning in the context of eyes and optics appeared in a few places in the first season of the *Daredevil* television show and has occasionally been hinted at in the comics as well. I am talking about the notion that Matt can sense sources of heat remotely, and do so in ways that suggest that these impressions can make a kind of image.

The first thing to keep in mind here is that the human skin works nothing like the film of a camera, and does not have any special ability to detect photons the way the retina does. This is true whether we are talking about visible light or the longer-wavelength infrared radiation which emanates from all objects in proportion to their temperature (a topic for chapter eight). The latter can be made visible to us by special cameras or goggles that translate these wavelengths into something we can see, but the emphasis here is on "see." As you will recall from our brief look at thermosensation, photosensitivity to infrared light is *not* the process by which the skin detects changes in temperature. Both infrared and visible light can obviously warm the skin, but this sensation is relayed by the various temperature-gated receptor proteins that detect warming and cooling, not by photoreceptors.

However, even *if* we were to go along with the idea that Matt's skin functions like a large retina that can detect infrared radiation, he would not be able to form an image this way. Instead, his skin would be like one giant thermosensitive eyespot. He might be able to tell the relative location of a particularly warm object, just the way *all* of us can, but his skin cannot form an image any more than a retina on your cheek can. In fact, the few animals in nature that *can* use thermosensation to detect the precise location of nearby prey, which include a family of snakes called pit vipers, are only able to do so by combining traditional thermosensation with specialized organs. What do these "pit" organs do? As it turns out, they form a cavity – not unlike a pinhole camera – to make a crude *image*!

To suggest that young Matt can detect that a woman passing by has skin that is "too hot," as in a scene from episode seven of the first season, is no less of a stretch than to suggest that he can literally see with his skin. I doubt that the show's writers had any real sense that this is what they were doing, but it underscores how difficult it can be for creators of every era to fully divorce themselves from the logic of vision.

One thing to ponder in all of this, when looking at a whole sensory system or organ, is that the failure of certain sensory receptors to respond to some incoming stimulus is as much a feature of the system as the presence of the sensory receptors that *do* respond. In fact, drawing a strict boundary between "here" and "not here" – and "this" and "not that" – is important enough that our higher-order sensory systems actively work to reduce the neural activity near the site of maximum activation. This process is called *lateral inhibition* and helps make for a cleaner signal that doesn't blur the boundaries unnecessarily.

There might be something to gain in overall sensitivity to sound, for instance, if all of our hair cells responded to a wider range of frequencies, but we would pay a price in the quality and "richness" of the sound. The pattern repeats itself with smell which also depends on very specific interactions between stimulus and sense organ. If the scent of a rose did not activate only a specific pattern of receptors in the olfactory epithelium – and *only* those receptors – then it would be indistinguishable from all the scents that are "not rose."

In *"Cut Man,"* the second episode of the *Daredevil* television show, there is a scene showing young Matt Murdock practicing his braille reading. While explaining to his father how it is done, he says "You have to pay attention to what isn't there, as much as what is." As we have just seen, this is a concept that applies to sensation more broadly. Equally important is the ability to tell the difference between different intensities of otherwise similar stimuli so that we can tell a softer sound from a more intense one, for instance. Being able to make these distinctions, when it comes to the world of sound, provides us with information about the location of sound sources, and their quality.

This may sound obvious to the point of banality but is important to keep in mind where Daredevil is concerned, as writers often have a tendency to want to turn every sensory dial to the max. For instance, in *Daredevil* #151, by Jim Shooter, Roger McKenzie, and Klaus Janson, a caption reads: "The words are *whispered*, but even through the closed door they boom like *shotgun* blasts to Matt Murdock's super-sensitive ears."

Of course, we don't know whether this is meant to be understood as sounding like a shotgun blast to Matt or to us readers, but this would not be the first time that it has been suggested that Matt can somehow hear a wide range of sounds equally well – similar claims are made throughout the history of the *Daredevil* comic – or sort through them as if they were different stations on a radio.

In reality, this would create more problems than it could ever hope to solve. The fact that a whisper sounds softer than a shotgun blast is one of the many things about this sound that makes it identi-fiable *as* a whisper. However loud a nearby whisper sounds to Matt Murdock, it should at the very least sound softer than a shout from the same distance. "Not hearing," "not smelling," and "not sensing," is just as important to Daredevil as it is to us, no matter how sensitive his heightened senses are.

WHAT IS MEANT BY "HEIGHTENED" SENSES?

Now that we have a grounding in how our senses work, we can start to explore what "heightened" senses might be like. Because it should be clear at this point that our senses can vary along more than one dimension, and "heightened" can mean more than one thing.

For instance, when people suggest that dogs hear better than us humans, what does that actually mean? My guess is that it's usually a simple reflection on the fact that their ears perk up at sounds we cannot detect. But what is it that makes them undetectable to us?

Well, it is not that they are too *faint*. You can take any of those sounds that are outside of our hearing range, but audible to a dog, and turn it up loud enough to make your poor pup whimper, and we would still not be able to hear a thing. To us, hearing a sound at, say, 35,000 Hz is equivalent to being able to see ultraviolet light. It cannot be done under any circumstances, because these stimuli lie outside the natural range of what our ears can detect.

However, dogs do not actually hear better than we do at all frequencies. And in this case, I'm using "better" to refer to the *sensitivity* of our hearing at frequencies we actually *can* hear. The absolute threshold of hearing gives us the intensity of sound at which said sound is just barely audible, and this measure varies greatly by frequency. Human hearing is, in fact, exceptionally sensitive, especially between the frequencies of roughly 500 to 4,000 Hz. There are also sounds that *we* can hear that dogs cannot, which we'll get back to shortly.

Another way to theoretically make our sense of hearing "better" is one we touched on in the previous section. Rather than extending its range or sensitivity, we could play with the idea of adjusting the equivalent of its "resolution." This would allow us to be better able to *discriminate* between the different frequencies we can hear, and between different intensities of sound. Such an enhancement would do little to allow Matt Murdock to hear heartbeats but could be very useful for making precise judgments about sound sources and echoes.

An additional factor to keep in mind about sound and hearing is that they are "temporal" phenomena. That is, sounds unfold over *time*. An image can be viewed as a stable two-dimensional sort of thing with clear spatial boundaries. Of course, this doesn't mean that images don't change; real scenes are obviously very dynamic with people and objects moving around. But it means that it is possible to take a snapshot of a scene from the visual world, and have the entirety of the image – *at that point in time* – available to us. Sounds cannot be understood this way.

One obvious example of this is speech, which we receive as a series of different sounds that are produced over the course of some measurable amount of time. Other natural sounds also have beginnings and ends and changes that happen in between. In order to be able to distinguish and recognize complex sounds, such as spoken words in any language, our ears and brains need to be able to give us sufficient *temporal* resolution. Think of a stream of sound like a loaf of bread on a conveyor belt about to go into a slicing machine. The rate of the conveyor belt is constant (i.e. the speed of sound), but the rate at which the blade chops that loaf into individual slices may differ. Your ears too, have a chopping rate (not a real scientific term), which also contributes to the overall resolution of hearing. This would be another factor to play around with in the pursuit of a "heightened" sense of hearing, or hearing of better *quality*.

What about touch? My guess is that we typically associate the idea of a heightened sense of touch with the ability to detect fine detail. As in the case of vision, this comes down to spatial resolution more than it does the absolute threshold of the touch receptors. We will delve into the case of reading print by touch in chapter eight, but you would need – at the very least – more of the relevant touch receptors, and for them to sit closer together and have smaller receptive fields if you are going to make a very significant difference here.

When it comes to the sense of smell, we are presumably once again back to absolute thresholds, with the assumption being that Daredevil can detect scents at lower concentrations than we can. Less has been stated about any ability to make fine distinctions between scents unless we think recognizing "any man by his hair tonic" is the epitome of Daredevil's olfactory prowess.

The subfield within the area of psychology that studies the relationship between physical stimuli out in the real world and the subjective sensations they give rise to is called *psychophysics*. All of the different ways of measuring the senses I've touched on here are things that this discipline concerns itself with. We have already acquainted ourselves with the so-called absolute threshold, which is one important property of the interaction between a stimulus and a sensory system. That is, at what *intensity* of any particular stimulus is there just enough for us to be able to reliably detect it?

Psychophysics also measures how well we perceive *differences* between stimuli, which is obviously an indication of the resolving power of various aspects of a sensory system. One central concept in this regard is the "just noticeable difference," often shortened to simply "jnd." For hearing, we might ask how different two sounds have to be in terms of *frequency* for us to be able to determine that they are in fact different sounds. Or, how big does the *intensity* difference have to be between two otherwise identical sounds in order for us to determine that one is softer than the other?

The size of the "jnd" in absolute terms is not constant across all frequencies and intensities though. In fact, whether or not you can detect a difference between two stimuli depends on the *relative magnitude* of this difference, and this is true to varying degrees for all the senses. This makes intuitive sense. If you add 50 grams (or about 2 ounces) of liquid to one of two empty and otherwise identical paper cups, you can easily perceive the difference in weight between them. Add the same amount of weight to one of two 8-pound kettlebells, and odds are you would find the same task quite difficult. The same thing goes for sensation generally.

Imagining what it is like to be Daredevil, and what having heightened senses might entail, requires that we play around with the different "settings" relevant to each sense, both in terms of ranges, absolute thresholds, and other measures that determine the quality – not just the "quantity"! – of the senses.

However, before we start playing around with that stuff, we may want to be aware of what kinds of rules we are breaking in doing so. What I'm getting at is that before we endow Daredevil with the ability to hear a dog whistle, it might be helpful to know why none of us can in real life, and exactly how far-fetched such a proposal would be. What does it mean to say that a sound at 35,000 Hz lies outside our natural range? And what would we need to do to bring it *inside* our range? What actually dictates the ranges we can hear, or which smells we can detect?

One easy answer would be "evolution," but that is really only half of an answer. Evolution is constrained by the fact that all life on the planet is made up of the same basic ingredients. We all use DNA to contain the recipes for building cells and proteins, and these recipes all draw from the same molecular ingredients in the form of twenty different amino acids which in turn consist of various constellations of common elements. These basic chemical components of life had to be present in sufficient amounts on our planet when life arose.

Evolution is also constrained by the hard physical realities that come with life on Earth. We have already touched on how physics constrains the stimulus. Similarly, it is to be expected that the evolution of our sensory systems is bound by the constraints that come with our specific environment. Our planet is of a particular size, with a particular force of gravity, and located at a particular distance from the nearest star. This star, our sun, has a particular range of energy output in the form of electromagnetic radiation. When it reaches Earth, the sun's rays are filtered by an atmosphere with a particular chemical makeup. As luck would have it, the same frequencies of light which are plentiful near the Earth's surface are mostly the same frequencies that can easily transcend the depths of the oceans where much of our earliest evolution took place. It is no coincidence that the evolution of photoreceptors, and then more complex eyes, has been shaped by the source of light that was plentiful in the oceans.

If you are the owner of a cat or small dog, take a look at the size of their eyes. You will probably find them to be slightly smaller than yours, but much larger when compared to the size of their heads and bodies. In the unlikely event that you are the owner of a pet elephant, you will find their eyes to be relatively *smaller* compared to their overall body size. Why is this? The answer lies in the physics of light and its interactions with an optical system, in much the same way as the physics of sound sets limits on the inner workings of the cochlea and other structures of the middle and inner ear. An animal's ability to capture a high-resolution image comes down not only to how well the eye operates along the lines of what we talked about earlier but also to the size of the eye.[5]

It turns out that the length of the eye, that is the distance between the pupil and the retina, corresponds with visual acuity in such a way that longer (bigger) eyes make for a larger retinal image. As for why elephants don't have bigger eyes than they do, if bigger is indeed better, we need to keep in mind that there are other evolutionary forces at play as well. Eyes are metabolically expensive to maintain, as are the areas of the brain in charge of processing the input, and an unnecessarily large eye is also more prone to injury. There is clearly a sweet spot here.

We can make similar observations about sound. The physics of sound – including the scale of the stimulus – is important enough that the dimensions of the structures of the ear of a given animal emerge as a compromise between the size of the animal, and this absolute physical scale. We see this very clearly in mammals.

All of this is obviously guided by evolution, but evolution itself is subject to physical realities. This is why you may be twenty times larger than your cat, but the structures of your outer, middle and inner ear will be much closer in size. The basilar membrane, which runs the length of the snail-shaped cochlea, and houses the hair cells which convert sounds into electrical signals, is only about thirty percent longer in humans than it is in cats. And the basilar membrane of an elephant is less than ten times as long as that of a mouse, despite the fact that the former weighs approximately 20,000 times more!

The topic of biological scale provides at least part of the answer to why humans can't hear dog whistles. Among mammals, there is a statistically significant relationship between the size of the animal and the frequencies the animal is most sensitive to. There is an even better correlation between this peak frequency and the length of the basilar membrane, but since the latter is, in turn, correlated with body size the model also holds pretty well for body size alone.

There is also a correlation between the physical distance between the ears of an animal, which obviously depends on the size of the head, and the upper limit of its hearing range. For reasons I'll get back to in the next chapter, the smaller your head is, the higher your hearing range needs to reach, in terms of frequency, if you are going to be able to use sound effectively to locate sound sources. It's another beautiful example of physics and biology coming together. All of what I have just mentioned help us account for why elephants have a lower frequency "peak" than humans. We, in turn, hear best at frequencies slightly lower than the frequencies at which dogs enjoy *their* best hearing.

In fact, most species of mammals tend to have hearing ranges that are fairly similar in *scope*, just shifted up and down. For instance, for sounds with an intensity of 40 dB, most mammals have a hearing range that covers between seven and nine octaves. Under these circumstances, both dogs and humans can hear across a range of about eight octaves, with the difference being that our hearing range starts lower, and theirs ends higher.

We usually associate the concept of octave with music theory, and pitch perception. For our purposes here, one octave simply represents a halving (one octave lower) or a doubling (one octave higher), compared to some reference "starting" frequency. For humans, the sounds a young person can hear at the relatively soft sound pressure level of 40 dB, start at around 60Hz. Eight octaves (doublings) then give us: 120Hz; 240Hz; 480Hz; 960Hz; 1,920Hz; 3,840Hz; 7680Hz; and finally 15,360Hz. Take the exact numbers at each end of the range we ended up with here with a grain of salt, but this is a half-decent estimate and an illustration of how this principle works.

This is also important to keep in mind when comparing species. To get a fair sense of how the hearing ranges of different animals compare to each other, it is better to look at how many "doublings" their upper and lower frequency limits encompass. Certain echolocating bats can hear sounds as high as 200,000 Hz. However, the *lower* end of their hearing range is only as "low" as around 10,000 Hz (which, incidentally, is where the hearing range of the elephant *ends*).

So, you could either say that their hearing range stretches an amazing 190,000Hz (from 10,000 to 200,000!), or you could say their hearing range stretches just over four measly octaves, well below that of many other mammals. It suddenly doesn't look quite as impressive, does it?

I am not saying this to dunk on the genuinely amazing hearing abilities of any particular bat but to stress that you need to understand what it is you are comparing. I cannot tell you the number of pop science Internet sites that confidently proclaim dogs, mice, and bats to be auditory superstars more capable than the poor, impoverished *Homo sapiens,* without taking into account the physical realities of sound and the different ecological niches we all occupy.

I am convinced that this is the kind of misunderstanding that would have us believe that human hearing is not particularly good compared to that of dogs, which objectively speaking isn't the case at all. We have the hearing apparatus you would expect of a fairly typical mammal of our particular size, which is a big part of why we can't hear sounds at 35,000 Hz but can hear relatively low sounds. Where human hearing really shines is in the later stages of auditory processing in the brain. We take our ability to understand the human speech of our surrounding culture for granted, but this is truly an amazing achievement, and much more relevant to our lives than hearing dog whistles.

That is not to say that there are no "auditory superstars" on the level of the ear itself for us to compare ourselves to. Or rather, for us to compare Daredevil to. Given that Matt Murdock is an otherwise typical human of typical size, it makes sense for his hearing "sweet spot," in terms of which frequencies he's most sensitive to, to be similar to that of typical humans, especially since this is shaped in part by the dimensions of the ear canal. But since the "miracle exception" offers us this one chance to actually meddle with the design of his middle and inner ears, no questions asked, what other animal might we seek to emulate, if we want to *expand* Matt's hearing range as opposed to just shift it up or down? While we can rule out our canine companions, there is another common household pet we can look to for inspiration – the domestic cat. Cats really do hear "better" than us, *and* dogs, by several measures. This, of course, makes their tendency to constantly ignore us all the more aggravating.[6]

I mentioned earlier that both dogs and humans have a hearing range of about eight octaves for sounds at a sound pressure level of 40 dB. The same range for cats is 9.3 octaves. An even more dramatic difference can be seen with *quieter* sounds, though. At a very soft 20 dB, cats have a hearing range stretching 8.1 octaves, or a whopping two(!) octaves more than dogs and humans. At a louder 60 dB, cats hear sounds between 53 Hz and an astonishing 77,000 Hz. Our hearing range begins almost an octave lower, at around 30 Hz at this sound pressure level, but it ends at 19,000 Hz, just over two octaves lower.[7]

But if dogs, cats, humans, elephants – and even bats – all rely on a common mammalian blueprint to guide the development of the cochlea and the rest of the auditory system, what allows certain animals to perform better than average? Well, one important factor may be the shape of the cochlea, specifically the number of spiral turns of this tiny snail-shaped organ. An increased number of turns appears to be a factor in extending the hearing range further down into the lower frequencies than would otherwise be expected from the length of the basilar membrane (which in turn is correlated with body size).[8] Genes also make for interesting targets of comic book meddling, and there is one in particular that I will return to in chapter five.

Going by James Kakalios's definition, the miracle exception covers the event and circumstances that allow someone to have heightened senses, and thus conveniently takes care of at least some of the physical challenges to the limits of human biology. However, as I hope I've been able to convey, even these aspects of Daredevil's power set are open to examination in really fascinating ways. We are going to continue this exploration in later chapters, beginning first with an organ I may have downplayed up until now: the brain.

The interactions between our sense organs and their respective stimuli obviously set limits for the kind and amount of information that will make it to the brain in the first place, which is why changes to the brain alone will not affect the gross amount of raw information that Daredevil, or any of us, has access to. The ways in which changes to the brain would make a difference in the realm of heightened senses is in how the information from the senses is processed. Is there anything we could do to squeeze a little more action out of this three-pound chunk of nervous tissue? Certainly!

First, though, we'll do a quick tour through the brain basics, because it should go without saying that none of us are able to consciously perceive *anything* without a sophisticated nervous system. And wouldn't that be a shame?

CHAPTER 4
SENSES, MEET BRAIN

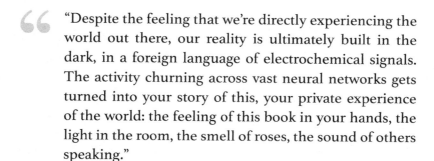

"Despite the feeling that we're directly experiencing the world out there, our reality is ultimately built in the dark, in a foreign language of electrochemical signals. The activity churning across vast neural networks gets turned into your story of this, your private experience of the world: the feeling of this book in your hands, the light in the room, the smell of roses, the sound of others speaking."

The Brain: The Story of You, by David Eagleman

At the heart of Daredevil's foundational concept as a character is that his remaining senses were heightened as compensation for his blindness. Young Matt Murdock was "helped" by the radiation that took his sight (in ways that only make sense in comics), but as we learned in chapter one, the notion that blindness brings gifts in other domains permeates cultures around the world. And while the stories told about blind characters in literary history are rooted more in folklore than science, sensory compensation as an idea hasn't appeared in a vacuum. Surely, there has to be *something* to it? The answer, it turns out, is yes! Maybe not enough to give anyone outright super-powers, but it is certainly the case that blind people – as a group and depending on such factors as the age of blindness onset – do outper-form sighted people on particular tasks. And the difference does indeed boil down to what the *brain* is doing, which is pretty darn cool.

Before getting into the science of the brain and sensory compen-sation, we should probably briefly address another factor that may have contributed to the myth that blind people have outright super-powers, which is that the average person typically has such low expectations of what a person who is blind can actually do that even the performance of mundane tasks might inspire awe from an unin-formed public. In his 1959 autobiography *The Sound of The Walls*, the writer and historian Jacob Twersky, who lost his sight in childhood, remembers the following from his time at a school for the blind in the 1930s:

"The public exhibition before the Easter vacation filled me with bitterness. We performed for seeing visitors, relatives, and others. Girls were shown at cooking. Boys were shown at gym work, including a tug-of-war. Boys and girls were shown reading and writing Braille. We were exhibited at a variety of tasks, none of which required uncommon ability, nor would seeing children have been exhibited at any of them. We were pitied by the visitors, marvelled at."

Even when it comes to the *Daredevil* comic, this is important to keep in mind. While we don't expect people with severely limited vision in our own world to go out and fight ninjas – I would person-ally advise against this regardless of a person's vision status – what Matt Murdock does in his civilian life as a lawyer, and otherwise "out of costume," requires no superpowers.

Just because those of us wholly unaccustomed to vision loss would feel quite helpless if temporarily blinded does not mean that this is the case for people with many years of experience with blind-ness, who may have developed a wide range of strategies and tech-niques for successfully negotiating their environments. If a skilled

blind person going about his day *seems* like he must have superpowers, or "heightened senses," that's our own lack of imagination talking. Vision is so central to most people's experiences that those of us
who have it cannot fathom being able to do much of anything
without it. But humans are resourceful creatures, which is yet
another thing we might thank our brains for.

THE ONE-NOTE LANGUAGE OF THE NERVOUS SYSTEM

We spent the last couple of chapters focusing on what sorts of stimuli
are available for us to sense, and how they interact with our sensory
organs through dedicated sensory receptors endowed with specialized protein micro-machinery. This first step, during which a proximal stimulus is detected and converted into a chemically mediated
electrical signal, is called *transduction*. The signal, or neural impulse,
which travels along the nerve fiber to the brain, is called the *action
potential* and is at once one of the most banal and fascinating
phenomena in nature.

The action potential is common to all the senses – all neural
activity, in fact – regardless of which sensory pathway gave rise to it.
Nothing about the nature of the particular stimulus from the outside
world is preserved in this spike of activity. In fact, the action potential
is so basic that we can liken it to a kind of morse code, but instead of
having short pulses and long pulses, there is just one sort of pulse.
The action potential is all or nothing. A more intense stimulus causes
more signals to be sent, and through more nerve fibers, but the
magnitude of the individual spike doesn't change.[1]

What does this neural impulse consist of? You may recall from
the previous chapter that the way for a sensory receptor to signal its
"on-state" is to open ion channels, which causes a drop in voltage.
The default so-called *resting voltage* across the membrane of a neuron
is -70 mV (millivolts). If enough ion channels open in response to a
stimulus to rapidly reduce the difference across the membrane to
around -55 mV, this will trigger the further opening of even *more* ion
channels. These particular ion channels are said to be *voltage-gated*

since they open or close in response to a change in voltage. These drastic changes in ion flow reverse the charge of the membrane potential which peaks at +40 mV. At this point, the voltage-gated ion channels have rapidly begun to close, which reverses the whole process. Remember that we also have those hard-working ion pumps to help put all the ions back where they belong when the membrane is at rest. In fact, this step actually overshoots its goal so that the potential reaches -90 mV before stabilizing back at -70 mV.

This entire cycle takes about four-thousandths of a second for any patch of the cell membrane to complete, but this little spike of activity doesn't stay in place, it moves from where it started, up the length of the axon (nerve fiber) like a lit fuse. This is how the action potential works, with the exception that the fuse can be reignited again almost immediately. The way for a lone sensory receptor, say a collection of free nerve endings equipped with temperature-gated ion channels, to signal "Hey, it's getting warmer over here!" is to first detect enough of the fun stuff going on to reach the threshold level of that first spike, and then another, and another. Activity is signaled by a marked uptick in the rate of firing of action potentials through the nerve.[2]

This is the extent of the sophistication of the basic unit of the neural code. It is like the language of the sentient tree-like life form Groot, known as a member of Marvel's *Guardians of the Galaxy*. Except that unlike Groot's single utterance, "I am Groot," it is not varied in inflection or force of delivery, but is instead pronounced the exact same way every time, just with varying rates of repetition.

Of course, the ways in which the nervous system is able to put the one-word language of the action potential to use is quite remarkable in itself. Neurons differ greatly in what kinds of networks they create with other neurons and the nature of these interactions. Neurons can make direct physical connections with other neurons by something called a "gap junction" which allows the action potential from one cell to continue to propagate in another. However, the most common kind of connection neurons make with each other is to form a *synapse*.

A synaptic connection will have a "pre-synaptic" sending neuron release a signaling molecule into the space between it and the "post-synaptic" receiving neuron. This tiny space is called the synaptic cleft, and the signaling molecule is called a neurotransmitter. The post-synaptic neuron is activated by the binding of the neurotransmitter to receptor proteins, which can then give the signal for continued activity, including triggering another action potential. Didn't I tell you that receptor proteins were kind of a big deal?

Among the neurotransmitters commonly known to the general public are serotonin and dopamine, but there is obviously a long list of these substances that are found more or less frequently in different parts of the nervous system. For instance, the neurotransmitter that the motor neurons use to talk to muscle cells, making them contract, is acetylcholine. Meanwhile, the most common neurotransmitter in the brain and spinal cord is glutamate. Glutamate is *excitatory*, which means that it contributes to the firing of an action potential in the target neuron. The most common *inhibitory* neurotransmitter in the brain is a molecule called GABA, which has the opposite effect.

Exactly how a particular neurotransmitter works on its target depends on what kind of receptor is expressed by the post-synaptic neuron. Among the receptor proteins that are activated by glutamate, for instance, we find both ion channel receptors and GPCRs, and it is ultimately the activity of the receptor protein that determines how the activity of the neurotransmitter is to be interpreted by the receiving cell.

So, how is it that we are able to have such diverse sensory experiences, to say nothing of all the *other* things we use our brains for if it all just boils down to spikes of electricity and neural chemistry? Even with the added layers of complexity afforded by the specifics of the wiring of the brain, and the way the activity of the neurons is modulated by different neurotransmitters, we are still dealing with a scenario where everything we perceive, and every thought we have, has to be assembled from patterns of electrical activity.

It remains difficult to comprehend, much less explain, how this process ultimately gives rise to the "feels" we associate with each sense, to say nothing of higher consciousness. The perception of C minor, the color red, and the bouquet of a baby's diaper could not be more different. This "what-it-is-likeness," of a sensory experience is known in philosophy as *qualia*. The neurological basis of qualia remains elusive. In fact, what has been called "the hard problem of consciousness" centers specifically on our difficulty in explaining and accounting for these different felt mental states.[3]

I will obviously not be making any attempts to solve this problem. There are scores of neuroscientists and philosophers of mind who spend considerable time thinking (and arguing) about the nature of conscious experience who are better equipped to do so than I am, and even they are far from reaching any kind of consensus. This doesn't mean that either sensory qualia or higher consciousness are inherently unexplainable, just that we still have a long way to go.

I must confess, however, that I find the topic fascinating, especially as it pertains to the understanding of Matt Murdock's imagined inner life. While we cannot (yet) explain why hearing "feels" like hearing, and vision "feels" like vision, we can at least establish that one is not like the other. And whatever the difference may be, it does *not* boil down to a difference on the level of the individual action potential. Something has to happen between the point of interface between the stimulus and the nervous system and the fully realized experience provided by the brain's processing. Is there perhaps something about how the primary visual cortex is wired that makes *anything* sent its way "vision-like?" This was one question the philosopher and psychologist William James posed in 1890 when he wondered whether crossing the wires from the eyes and the ears to their respective centers in the brain would lead a person to "hear the lightning and see the thunder."[4]

The results of a series of groundbreaking experiments, beginning in the late 1980s in the lab of Mriganka Sur at MIT, may help us answer this question. By surgically redirecting some of the nerve fibers coming from the eyes of newborn ferrets to instead project to the auditory pathway, the experimenters found that a pattern emerged in the auditory cortex of the ferrets that made it look and act strikingly like what you would expect from the *visual* cortex. Contrary to what you would assume from its original "programming," and to James's suggestion. The rewired auditory cortex was not only able to appropriately accommodate the new signal, but it also allowed the ferrets to actually *see*, and to experience this input as being visual in nature. A flashing light shown to the rewired part of the visual field did *not* cause these ferrets to hear a sound.[5]

What you *might* take away from these findings is that the different areas of the brain are not, in fact, specialized to receive a particular type of input. This would be inaccurate. The auditory cortex of the rewired ferrets was not quite as good at this new job as you would expect from the visual cortex. The blueprint for building a brain (ferret or human), complete with the appropriate wiring, is contained in our DNA, and it makes sense that different brain regions would be optimized for their expected tasks, and develop to be at least partially pre-assembled. Instead, the takeaway should be that the neural pattern of the input is *one* key driver in telling the brain what kind of a "thing" this input happens to be, and how it might best be processed. What made the rewired auditory cortex of the ferrets come to "perform" vision was that the information being received conformed to a kind of visual "logic."

A single neuron firing isn't cause for much excitement, but a whole choir of "Groots" can generate particular patterns across both time and space (such as the cross-section of a bundle of nerve fibers) in ways that vary from one sense to the next, based on the contingencies between things and events in the environment and the animal moving through it.

PATTERNS IN THE BRAIN

What does it mean to say that the rewired auditory cortex of the ferrets at MIT had started to look and behave similarly to the visual cortex? What would such a pattern look like and what does it have to do with how we perceive the visual world?

What ordinarily happens when a person sees something through the eyes, using a presumably unadulterated brain, is that the input from the retina is sent along various routes. Some of the "stations" inside the brain do basic yet important things like regulating the sleep-wake cycle, the size of the pupil, and certain types of eye movements.[6] The main highway, however, leads to a structure inside the thalamus called the lateral geniculate nucleus (LGN), and then on to the primary visual cortex (V1), also sometimes referred to as the striate cortex (based on its striped appearance).[7] Of course, there's some crossing and uncrossing of wires along the way so that the information from the left visual field gets sent to the right hemisphere of the brain and vice versa.

What you will discover when you look at the neural structure of both the LGN and the V1 is that the "topography" of the retina, itself a two-dimensional spatial map of the visual world, is largely maintained. This kind of representation is said to be *retinotopic*. Each spot in the visual field is detected by a particular patch of the retina, which in turn sends its information to a particular cluster of neurons in the LGN and then on to a corresponding cluster in V1. If two spots are right next to each other in the visual field, the information from these spots will be processed in locations right next to each other in the brain as well. *This* was the kind of map, along with other typical features of the microcircuitry of the "visual brain," that developed in the auditory cortex of the rewired ferrets in response to things they saw.

Retinotopy remains a feature of other visual areas beyond V1, but the maps may pick up various distortions along the way, depending on what kinds of things about the scene these areas are interested in. In fact, the map starts out distorted too, as the central visual field has much more neural real estate dedicated to it than does the periphery.

One thing that makes vision such a difficult nut for the brain to crack is that it cannot simply take in *all* the information from a particular location in space as if it were a movie screen. Remember, all the brain is getting from the eyes is electrical activity, and everything a person sees is literally in their head. The visual scene needs to be broken down into its tiny constituent parts and features by neurons with exceedingly narrow job descriptions, only to be sewn together, through subsequent (and parallel) steps of neural processing into something that we perceive as a seamless whole. Since individual neurons are not equipped to answer essay questions, they have to deal largely with 'yes' and 'no.' A particular neuron might fire only if its dedicated spot has a line or edge in it at a particular angle. Or if it detects a particular color. Some neurons may deal with only information coming from one eye, while others piece together information from both eyes. The processing of other kinds of sensory information also has to proceed along similar steps.

What, then, would we expect of the auditory system? Information from the cochlea to the primary auditory cortex (AI) goes through several more steps of processing than vision does on its way to VI. Along most of these steps, the information from the ears will be organized *tonotopically*. What does this mean? Well, if we think of the structure of the inner ear as a highly intricate xylophone with the higher frequencies on one end, and the lower frequencies on the other, we get a gradient from high to low (more on the inner ear in chapter five). The neurons at each station along the way up the chain are organized according to the same pattern, in bands of nerve cells corresponding to a gradient of sound frequencies.[8]

While the merging of the images from the two eyes happens in the primary visual cortex, the sound information received by both ears is integrated relatively early in the auditory system, by a structure in the medulla called the superior olive (there's a great name for a superhero). The superior olive on each side of the brain receives information from *both* ears. This makes it crucial for sound localization which is made possible by the fact that there is a short time lag between when a sound arrives at each ear if it is arriving from off-center. Depending on the sound frequency, there may also be a difference in sound intensity at each ear. As you will recall, lower frequencies with longer wavelengths easily bend around objects the size of a human head and won't lose much intensity, whereas our noggins more easily reflect sounds at higher frequencies (with wavelengths smaller than the size of the head). Whether the so-called interaural *time* difference (ITD) or the interaural (sound) *level* difference (ILD) matters more then depends on the particular sound. This is also why animals with small heads need to rely on higher-frequency sounds in order to make use of the ILD. The ITD is already less useful when the head is small since there won't be much of a difference to detect.

Just like the visual system, the auditory system has to disentangle many different properties of sounds, such as intensity, location, temporal patterns and much more. That we experience sounds as these fully formed, externally located "event-objects" is again a product of what the brain is doing.

What about the particular "topography" of the other senses? Most people, whether they can name him or not, are familiar with the work of American-Canadian neurosurgeon Wilder Penfield (1891-1976). Penfield pioneered the procedure for treating severe epilepsy by destroying the part of the brain where the seizures originate. During surgery, patients would remain awake while Penfield electrically stimulated the brain tissue around the site suspected of triggering the seizures. This presented him with a unique opportunity to conduct mapping studies of the brain, and he is most famous for the work he did on the motor and somatosensory cortices, or the parts of the brain devoted to movement and touch.

One representation of his findings is the illustration of the "cortical homunculus," an oddly assembled gaping man with a large head, and even larger hands, frequently shown wrapped around the side of one hemisphere of the brain.[9] The hands and face, especially the mouth and lips, are massively overrepresented. Like vision, the sense of touch then has a very obvious spatial component, except the surface on which a stimulus is detected is not the retina, but the skin. Penfield's homunculus shows that the brain maintains a representation of our bodies in such a way that a particular patch of skin, when touched, causes activation of a particular patch of neural circuitry in the primary somatosensory cortex (S1). This representation is then said to be *somatotopic*. "Soma" comes from the Greek word for "body," so we are very literally dealing with a "body map."

The topography of the olfactory system appears to defy any obvious logic in the sense that we cannot organize it spatially or according to any other kind of gradient. This probably isn't too surprising given our great difficulty in classifying scents even on an intellectual level. The chemists who have tried to do this using knowledge of molecular structure have also failed miserably.

Instead, the way we can think about the activation of the olfactory system is by imagining that the receptors found in the olfactory epithelium project to areas in the brain where they are arranged in a somewhat arbitrary "grid" where each kind of scent activates a particular pattern that is basically *combinatorial*.

Each specific olfactory receptor is typically activated by many different odor molecules. Each odor molecule, in turn, activates many different receptors. What makes each scent unique is the combination of the activation it elicits. The brain also responds differently to mixtures of scent molecules than you would expect from the combined pattern of each of the constituent scents presented separately. This is yet another reason why separating the scents of over a hundred murder victims seems like a bit of a stretch for even someone like Daredevil to do!

When I think of the encoding of odors by the brain, the image that comes to mind is that of a toddler randomly striking the keys of a giant church organ. (With the caveat that it's not random, and that the toddler probably needs more than two hands. Make it a baby octopus. Or maybe a centipede?)

· · ·

This way of presenting how sensory information is delivered from the periphery to the brain is by necessity overly simplistic. While it's appealing to think of it as a flow of information going in one direction, and from simple to complex, that's not really it. There is constant crosstalk between different regions of the brain, as well as feedback of information going the "wrong" way. V1 projects almost as many nerve fibers back to the LGN as it has coming in.

It's also tempting to think of the primary sensory cortices doing the "simple" stuff, and then pushing this up the chain for adjacent areas of the brain to extract ever more abstract information. But that is also overly simplistic. It is true that we find regions of the brain that are associated with very specific tasks. If you follow this field at all you may have heard of the two-stream hypothesis, according to which visual processing (in particular) is divided into a "where" pathway and a "what" pathway which take slightly different routes. The "where" pathway deals with the presence of objects in space and also supports actions toward them, whereas the "what" pathway supports the conscious recognition of objects. Likewise, damage to various areas of the brain can cause oddly specific impairments ranging from prosopagnosia (face blindness) to akinetopsia (the inability to detect motion), and cortical color blindness. However, despite this apparent "modularity," it is important to remember that no area of the brain works in isolation.

The last few years have seen an increased focus on not just figuring out which areas of the brain are active during particular tasks, which is something that has occupied cognitive neuroscience for much of the last thirty-odd years, but how they are all connected. The totality of all of the connections in a brain has been given the name "connectome."

So far we have but a single such connectome fully figured out, that of beloved model organism *Caenorhabditis elegans*. This critter is a tiny transparent roundworm, a mere one millimeter in length, with a brain consisting of 302 neurons, with 7,286 connections between them. The challenge of repeating this for the human brain, to say nothing of a particular individual (we are a more diverse lot than the roundworm), is that we have approximately 86 billion neurons sporting more than 100 trillion connections![10] Capturing the entire blueprint of the human brain is going to take a while, to put it mildly, and is currently beyond what we can reasonably accomplish. (I doubt anyone reading this in 2022 will be able to upload their consciousness to the cloud within the next fifty years.)

Of course, while knowing about every back alley and shallow footpath that connects small networks of neurons is currently beyond reach, we have an increasingly clear idea of the broad strokes; the freeways and state highways that connect major hubs of activity in the brain.

THE METAMODAL BRAIN

We just looked at some of the ways in which the brain processes the input from the different sensory organs. But what is the nature of the specialization that exists in the typical brain? If we look at the visual cortex, we can see that input from the eyes typically makes up the majority of its information diet, at least for people with a standard set of working eyes. But is this specialization for the visual channel restricted to that channel *per se*, or does it also support the processing of information from the other senses if the content and packaging of the information come sufficiently close to fitting its usual job description? In an influential 2001 paper, neurologists Alvaro Pascual-Leone and Roy Hamilton hypothesize that:

"the brain might actually represent a metamodal structure organized as operators that execute a given function or computation regardless of sensory input modality. Such operators might have a predilection for a given sensory input based on its relative suitability for the assigned computation. [...] In this view, the visual cortex is only visual because we have sight and because the assigned computation of the striate cortex is best accomplished using retinal, visual information."

What this means, in plain English, is that the visual channel would be better positioned to relay information about things like two-dimensional patterns or spatial relationships between objects than the other senses, and this is why visual brain areas specialize in dealing with these things. However, vision is not the only pathway we can use for accessing this information, and the brain might ultimately be structured around these "deeper" tasks and less concerned with through which channel the information was delivered.

There are, of course, sensory experiences that truly are "unimodal," in that we cannot conceive of them as existing outside the purview of a particular sensory modality. A scent, a tickle on the skin, or the audible quality of pitch are the sole responsibility of smell, touch, and hearing respectively. A source of sound may vibrate at a frequency that can also be felt by touch, but the experience of this frequency as having a particular *pitch* is as unique to the sense of hearing as the experience of color is to the sense of vision.

But what about a quality such as the shape of an object? I can see an empty glass in front of me, and in doing so will know its shape. If I close my eyes and reach out to touch it, my hand will also quickly recognize it by its shape, and both senses give me a very similar idea of its dimensions. The same often goes for motion. If I'm standing on a sidewalk, facing the road, and a car drives past me, this is something I can both see and hear, independently. The experience of tracking the car with my eyes may be quite different from tracking it with my ears, but the tracking *itself* can be done with a great degree of fidelity both ways.

Over the last couple of decades, we have seen evidence mount in support of "the metamodal brain." This perspective not only helps explain many of the phenomena we are about to encounter later in this chapter, such as sensory compensation and substitution but even some of the perfectly normal goings-on in the brain. We have known for some time that brain regions that are specialized for certain tasks appear to be pretty agnostic when it comes to the channel through which they get their raw information. Both spoken and signed languages activate the same major language areas of the brain for instance. Although this is perhaps less surprising when it comes to these more abstract abilities such as the processing and manipulation of linguistic symbols.

However, more recent findings suggest that even the *early* stages of sensory processing, those that typically deal with the "raw" information from the senses, may be less particular about where the information comes from than previously thought. A recent study comparing blind expert echolocators to blind and sighted controls found that the first group, but not the controls, showed retinotopic-like activation in the primary visual cortex in response to sound, and in particular to echo sounds![11]

It is also well-established that reading braille activates visual areas in (blind) braille readers. While subjects who have been blind since birth or early childhood show stronger activation that encompasses more areas, there is an effect for braille readers who have lost their sight later in life as well. These findings are not limited to braille but can be seen in the touch discrimination of embossed roman letters as well.

The activation of the visual cortex for reading braille is not simply a helpful complement to what the somatosensory cortex is doing. In skilled braille readers, the visual cortex takes on such an integral role that it becomes *necessary* for the task. Transcranial magnetic stimulation (TMS) is a non-invasive way to temporarily disrupt the normal function of select brain areas. When applied to the visual cortex of blind braille readers, they become significantly less accurate. Subjects will report distorted perceptions of "phantom" dots, and missing dots. There is also a case report of a congenitally blind woman who had been an avid braille reader since childhood until suffering a stroke to her visual cortex that left her completely unable to read. She simply couldn't make sense of the characters she was feeling, even though her touch sensitivity wasn't impaired.[12]

You could make the argument that the visual cortex of blind people is just looking desperately for something to do and that none of this necessarily supports a metamodal model of the brain more generally.

Does any of this apply to the brain of the average person with a full set of the typical human senses? Well, this is where things get even more intriguing. It turns out that delegating the task of braille reading to the visual cortex is not just for the blind; this division of labor arises spontaneously in the brains of the sighted as well!

Two studies of the same group of sighted subjects demonstrated differences in both anatomy and patterns of brain activity, before and after they had completed a nine-month program to learn how to read braille by touch (some of the subjects already knew how to read braille visually). At the end of this period, reading braille had started to activate certain areas of the visual cortex, as well as an area called the *visual word form area*, that is also active when sighted people read text visually. There were also noticeable changes in the anatomy of the visual cortex. What the experimenters found was increased connectivity between the somatosensory cortex and a part of the left primary visual cortex which collects input from the visual periphery. This area also grew in size.[13]

Worth noting here is that while these connections became stronger, they were also *already there* at the start of the experiment. The typical brain, residing in a body with all senses intact, apparently sees fit to have a decent amount of wiring between the areas dedicated to touch and the one that registers peripheral visual input. Apparently, there is something about the task of reading braille that the "visual" brain considers familiar enough that it readily volunteers to do it!

The studies referenced above involved a long learning process of several months. Another interesting study looked at the effects of just five days of blindfolding and gives us an indication of just how rapidly the brain can change, and how depriving it of visual input might help speed things along. The design of the experiment, which was conducted at Harvard Medical School over the course of three years, was quite robust, with a total of 47 subjects randomly assigned to either be blindfolded around the clock, or serve as non-blindfolded controls. The controls actually wore a blindfold during testing and training, which lasted several hours a day and consisted of formal braille instruction, playing tactile games, and some instruction on how to navigate with a cane. Subsets of both groups underwent testing of their tactile acuity, their recognition of braille characters, fMRI testing during a tactile task, and testing with a TMS device applied over the visual cortex.

As you might imagine, a test of the subjects' ability to recognize braille characters showed marked improvements in both groups over the five-day period. This is not surprising; practice makes perfect, as the saying goes. There was, however, a statistically significant difference between the performances of the two groups on day five, with the subjects who had been blindfolded around the clock for several days outperforming the control subjects. In terms of tactile sensitivity more generally, the tests performed showed improvements in both groups over the course of the study, with a slight edge for the blindfolded group, though not enough to reach statistical significance.

What happened in the *brains* of the blindfolded subjects during the course of the study is perhaps the most important, and frankly stunning, finding. The fMRI exam from day three showed little difference between the groups during tactile activation. By day five, however, there was statistically significant activation in the occipital lobe (which is home to most of the visual cortex) in the blindfolded group. TMS applied to this region on day five also created a significant increase in the error rate of braille character recognition in blindfolded subjects only. By the next day, though, after the blindfold had been removed, this effect was already gone, along with the activation of the visual cortex revealed by the fMRI.[14]

One important question that these findings help us address is whether the connections between different sensory areas of the brain that we see in people who are deaf or blind come about as a result of *new* connections or a result of the so-called "unmasking" of connections that were already there.

This last study offers substantial support for the latter interpretation. It seems unlikely that the brain would grow entirely new connections on such a scale in a matter of days, nor that it would let them fall idle the next day. These findings do not rule out that there could be more going on to explain the substantial changes that take place with long-term blindness (or deafness), especially in the case where it is present from birth or comes about early in life.

THE BRAIN WITHOUT SIGHT

Most of the studies into the general phenomenon of sensory compensation have been conducted with people who are totally blind. The focus on this particular group makes a great deal of sense, of course. For those of us with a standard-issue set of human senses, vision stands out as the most important one, and as much as fifty percent of the brain is involved in visual processing to at least some degree. Considering the information processing capacity of the human brain, and the fact that it gobbles up 25 percent of our energy budget, it would be an awful waste to not put this neural real estate to use. However, compensation phenomena are also seen in the deaf and have been extensively studied for this population as well.

Before going further into this topic, we should note that the kinds of changes you expect to see in the brains of blind individuals necessarily follow from the basic blueprint of the human brain, which is encoded by our genes. People born blind cannot be compared to the surgically rewired ferrets we met earlier in this chapter. Their brains have instead come to adapt to unusual circumstances naturally. This adaptation is made possible by the fact that the brain is naturally (though not infinitely) plastic. There are also, as we saw with the sighted subjects, established connections between and within the different sensory areas that can act as a stage for this process, where some connections will strengthen and others weaken. The young brain is also much more plastic than its fully matured counterpart, but the brain continues to change throughout life. If it did not, it would be very hard to pick up new skills or make new associations and memories.

.   .   .

What kinds of changes, then, do we expect to see when looking at the brains of people who are blind or deaf? On the anatomical level, the parts of the brain that receive input from the *remaining* senses, show signs of what is called *intramodal* plasticity. These parts of the brain don't start working outside their basic job description, but have to work a little "harder."

In people who are deaf, there is actually an expansion of the retinotopic map in the visual cortex so that more neural resources can be devoted to the visual periphery. Similarly, there is an expansion of the tonotopic map in the auditory cortex of blind people. This expansion is exactly what it sounds like – there is a representation of the same frequency ranges as in the average person, but they take up a larger area of the brain. This presumably allows the auditory cortex to devote more neural real estate to processing these sounds. Interestingly, this "map expansion" is evident both in subjects born blind, as well as in subjects who lose their sight after puberty.[15]

We see further evidence of intramodal plasticity in the somatosensory cortex of blind braille readers, such that the area of the cortex that is devoted to the main reading finger expands. An interesting finding concerns braille readers who use multiple fingers to read. This reading technique allows for more of the word to be read at the same time, which is good for reading speed, but has been found to cause a strange phenomenon: the brain begins to consider these fingertips as part of a whole unit. When asked to indicate which of their reading fingers has been touched, such subjects may be unable to answer![16] This shows that changes to the brain that helps a person accomplish one task may negatively affect some other ability.

However, with so much of the brain normally devoted to vision, it is in the realm of *cross-modal* plasticity that things get really interesting. The term cross-modal refers to the phenomenon where the areas that would normally be devoted mostly to the missing sense get recruited for other tasks.

Given the complementary research on sighted people and the concept of the metamodal brain, there appears to be some underlying logic that is driving the visual cortex to take on the task of braille reading, aside from its sheer availability. However, there has been some dispute in terms of whether this is the visual brain "performing" touch, as opposed to the visual brain "performing" language. Since reading braille is as much a linguistic task as it is a tactile one, these effects may be hard to disentangle, and some overlap is to be expected.

It may sound odd to expect the visual cortex to be involved in language, but this is very much the case in blind people! One study tested sighted and blind (early and late) subjects on a task in which they had to generate a suitable verb to go with a heard noun. The groups showed similar activation of the typical language areas you would expect, but the blind subjects showed additional activation in areas of the visual cortex known to be active in braille reading. Interestingly, the activation of these areas in the blind was localized to the *left* hemisphere of the brain, similarly to other language areas.

The dominant theory is that the visual cortex is well suited to help with the cognitively more demanding tasks of language processing, such as semantics (the meanings of words and sentences) and grammar. This would also be consistent with the finding that the visual cortex is involved in verbal memory in the blind, showing activation when subjects remember lists of previously heard or read words. The blind as a group have also been found to outperform sighted subjects on such tasks, and their performance is directly correlated with the amount of activation of the visual cortex. Clearly, the involvement of these unusual additional brain regions is not just incidental, but aids directly in the performance of language tasks. Why is this useful? One theory is that an inability to see translates into a stronger reliance on information conveyed by language. Rather than simply looking at where someone is pointing when asking for directions, for instance, the blind person may need to store this information as a longer description with a series of steps to be remembered.

Blind people who use screen-reading technology with synthetic speech output are also able to decipher this speech at rates that frankly boggle the mind. Untrained subjects can decipher about eight syllables per second, whereas blind experts can understand speech at up to 22 syllables per second! This too appears to involve the primary visual cortex, specifically the right hemisphere, in conjunction with other brain areas.[17]

With language providing a nice segue from a discussion of anatomy to one of task performance, let us look at the complicated case of the sense of hearing in the blind. So far, we may have been led to believe that losing a sense may bring nothing but benefits when it comes to the spared senses – just look at Daredevil! But this is not in any way a given; nor is it a universal truth.

The young brain that has access to information from both vision and hearing can use the visual channel to help calibrate auditory space. If a young child sees a bouncing ball and simultaneously hears the sound of the bounce, this helps her brain map the exact location of the sound to where the ball was seen. This learning step is missing in people who were born blind or lost their sight early in life. This, in turn, helps explain why the congenitally blind in particular actually show deficits in spatial hearing on some tasks.

Determining the distance to a sound source proves to be a greater challenge to this group than to the sighted. The ability to determine the location of a sound source in the vertical (up-down) plane is also negatively affected by early blindness. Where the blind tend to perform better than the sighted is in determining the source of a sound in the horizontal plane, as in when a sound is coming from the side.

Interestingly, some blind people have also demonstrated an enhanced ability for horizontal detection of sound using just *one* ear, which in turn must depend on being able to make good use of spectral cues. We also use such cues to determine the location of a sound in the vertical plane (more on this in chapter five). Oddly enough, one study found that those blind individuals who performed the *best* on a monaural (one-ear) horizontal sound localization task, performed the worst at locating sound in the vertical plane, suggesting that this may result from a kind of trade-off that the brain has to make.[18]

One absolutely fascinating finding concerns the role that skilled echolocation may play in off-setting some of the negative effects of blindness on spatial hearing, and this is a topic I will return to in chapter six. For now, I will just briefly point out that while echo detection in general is another example of something the blind as a group tend to perform better at than sighted people, skilled use of echolocation is not something that develops spontaneously in a person just because she happens to be blind.[19]

On a final note, you may wonder what the research record has to say about any known compensations associated with blindness in the area of olfaction. Here, the findings are mixed and largely inconclusive. A recent systematic review of the literature found no systematic differences on metrics such as odor detection thresholds, and odor discrimination and identification.[20] However, one study found slightly better performance in early blind subjects on a measure of odor *localization*.[21] While humans have some capacity for "spatial smelling," this is not our best event (made more difficult by the nature of the stimulus), and seems largely dependent on some odors stimulating the trigeminal nerve, in addition to the olfactory system.[22] In general, it should be noted that the ability to put the sense of smell to good use is highly amenable to training.

SENSORY SUBSTITUTION

If sensory *compensation* describes the natural range of functional and anatomical consequences of a brain's adaptation to being a sense short, then sensory *substitution* typically goes a step beyond that.

The Encyclopedia of Neuroscience describes it as "[a] mechanism whereby information from one sensory modality is replaced by or substituted for information from another sensory modality," further adding that "[t]his term often refers to prosthetic devices that are meant to convey one form of sensory information through a sensory system that is typically not used for that form of information."[23]

The more general description is wide enough to potentially include echolocation to fill in for sight, and we'll get back to whether Daredevil's enigmatic "radar sense" might fit the bill. However, let us begin with the stricter definition which describes the use of some technical device to translate the patterns of one sense, typically sight, into a format that can be received by some other sense. For the substitution of vision, this would be either touch or hearing.

The field was more or less founded by Paul Bach-y-Rita, who in the late 1960's made the radical claim that we see with the brain, not the eyes.[24] The first experimental design that was tested by Bach-y-Rita and his team entailed hooking up a camera to a dentist's chair fitted with 400 vibrating pins in a 20x20 grid pattern. With the camera creating a grayscale image, the pins would press into the backs of blind subjects for the lighter pixels, creating a pattern on the back. The subjects were able to independently operate the camera to pan and zoom, and underwent extensive training to learn to differentiate and locate objects based on the images projected. After many hours of practice, they were able to do so quite reliably.[25]

This first system obviously proved difficult to take into the real world, but later devices have been more portable. For instance, a device called the BrainPort, developed by a company founded by Bach-y-Rita, miniaturizes the technology of vibrating pins by having the sensation delivered to the tongue via electrical stimulation. The bulky camera is replaced by one that is mounted to a pair of glasses. One famous user of the BrainPort is blind mountaineer Erik Weihenmayer, who uses it for rock climbing, among other things.[26]

Other devices make use of the sense of hearing rather than the sense of touch. The vOICe platform encodes visual information into a sound stream by scanning visual images from left to right, converting them into grayscale, and then dividing the image into pixels which are in turn converted into a sound with an intensity that correlates with the brightness of the particular spot. Pixels at the top of the image translate into sounds that are higher in pitch while sound in the left visual field is presented before sound in the right visual field during the second it takes for the device to finish one sweep. Other sound-based devices may incorporate more of an enhanced echolocation approach.

There are practical challenges to these devices, and using them effectively takes a great deal of practice. While something like the BrainPort has very poor *spatial* resolution, not far beyond Bach-y-Rita's dentist's chair, it has the benefit of being able to update quickly as the user moves around and thus offers a reasonable amount of *temporal* resolution. The vOICe and similar technologies have other challenges. While able to deliver more spatial detail, it also takes a second for each "scan" to finish which means that images are not delivered instantaneously and the temporal resolution is thus poorer than with a device like the BrainPort. An additional shortcoming with both technologies is that they do not automatically provide a sense of depth.[27]

Does sensory substitution actually "work"? Well, users can typically learn to recognize patterns and objects with practice, and in that sense these technologies can be useful, even though they have so far tended not to be widely adopted for everyday use. Getting back to a more philosophical perspective of sensation, we may want to ask what this is actually like. What kind of "qualia" does this give rise to? After all, a person could direct such a device at the letter 'E' written on a whiteboard and extract that information without having the sensation of actually *seeing* the 'E.' This would be a case of "coming to be informed about" rather than a perceptual experience that is similar to vision.

Looking at this aspect of the research yields an intriguing mix of outcomes. However, it seems that at least some subjects do perceive the object they sense as being out there in the space in front of them. The paper detailing the original study by Bach-y-Rita notes that "Our subjects spontaneously report the external localization of stimuli, in that sensory information seems to come from in front of the camera, rather than from the vibrotactors on their back."

Another 1974 paper by a G Guarniero, a congenitally blind doctoral candidate in philosophy, details the personal experience of spending three weeks learning to use a device called the TVSS, which is similar to the one developed by Bach-y-Rita, though it allows a little more mobility. Guarniero notes that "[o]nly when I first used the System did the sensations seem as if they were on my back. Later on, the sensations appeared to me to be in a two-dimensional space which did not seem to be 'out there'." Despite not experiencing the perceived objects to be in an external location, Guarniero very clearly found the experience to be quite profound, also noting, "Although the somatosensory cortex, only, was involved, nevertheless the experienced quality of the sensations was nothing like that perceived by touch. Rather than coining a new word, or using 'touch' or 'feel', I shall use the word 'see' to describe what I experienced."[28]

We now know that Guarniero was wrong to discount the visual cortex here. Sensory substitution devices, when used by trained subjects, do activate many of the same areas we are familiar with from the sensory compensation literature, including the visual cortex. This is not surprising considering that these technologies take advantage of the same cross-modal connections already available in the brain.

While the nature of sensory substitution remains a topic of discussion, it raises some truly interesting philosophical questions. What is it really to see and can the experience of something similar to vision arise through other channels? Provided that we approximate the "logic" of vision in the pattern of information provided through this other sense, and allow for the proper exploration of the world so that the same sorts of contingencies are in place, the answer seems to be a solid "sort of."

You might wonder what bearing all this has on our understanding of Daredevil. I would say quite a bit. If we run the Matt Murdock thought experiment in our heads, it is easy to imagine that all of the neural processes that allow for sensory compensation and substitution to work their not-quite-magic are in place for him too. We can even happily bestow him with a little bit of extra plasticity if that helps. In addition, and looking at the way the perception of objects in external space can arise with the aid of sensory substitution devices, we can even make some predictions about what it is like to radar-sense.

We will spend the final third of this book looking at this "sense" from every conceivable angle, and I will also make a case for why echolocation, which might even be described as a natural form of sensory substitution, makes for the most elegant explanation of this Daredevil staple. However, some things must hold true no matter what we make of Matt Murdock's ability to detect silent objects in space. When reading through the *Daredevil* archives, it seems clear that one property of the "radar sense," at least according to most descriptions is that it also allows the title character to perceive objects in three-dimensional space, that is as *externally located*. And it is also a pretty safe guess that it is Matt's visual cortex that supports this and many other processes, in ways that are not unlike that of real-life blind people.

Before we get that far though, we will delve into the second part of our journey together by looking more closely at Matt's heightened senses of hearing, smell and touch. (Nope, we're not covering taste in a separate chapter until Marvel comes up with something more interesting than grains of salt on a pretzel!)

PART TWO
SUPER SENSES

CHAPTER 5
SUPER HEARING

 "There is no such thing as silence. We are constantly immersed in and affected by sound and vibration. This is true no matter where you go, from the deepest underwater trenches to the highest, almost airless peaks of the Himalayas. In truly quiet areas you can even hear the sound of air molecules vibrating inside your ear canals or the noise of the fluid in the ears themselves."

The Universal Sense: How Hearing Shapes the Mind, by Seth S. Horowitz

Having a heightened sense of hearing really is a great superpower. More so than vision, hearing allows us to be vigilant. Which is great if you're a *vigilante* – whether sighted or blind. Sounds can be heard from behind, and from the other side of a wall or objects that would obscure the line of sight. Hearing is also the fastest of our senses. The brain is much quicker to react to sound than to events in the visual field, and can also detect sound *changes* that unfold in less than a millionth of a second.

For a character like Daredevil, hearing stands out as perhaps his richest source of information. If we discount Daredevil's ill-advised adventures in piloting (more on this in chapter nine!), we can assume that distant sound sources represent the only stimulus that can bring Matt information from farther away than his immediate vicinity.

In the same book from which I've borrowed the introductory quote to this chapter, neuroscientist Seth S. Horowitz points out that when you pool together all the sources of sound that make up the background noise of an urban environment, you will be hearing sound from as far away as a quarter-mile (or 400 meters) in all directions. Most everyday sounds don't travel nearly this far however, at least not through the air, but vibrations also move through structures such as concrete buildings and asphalt-paved roads. The greater speed of sound through the ground is what allows particularly low frequencies to travel quite far.

Despite the fact that our auditory world is respectable in size, it cannot be compared to the visual world in this respect. This makes for an interesting difference between Matt Murdock and those of us with sight. On a clear day, and with nothing obstructing our path, we can see all the way to the horizon. We can even see the light from stars across the galaxy! Daredevil fans will sometimes poke gentle fun at Matt Murdock's narrow focus on the Hell's Kitchen neighborhood he calls home, but even if we account for super-hearing, Daredevil's blindness means that he is necessarily dealing with a smaller physical realm than many other heroes, and it actually makes sense that he would direct his focus to people and places relatively nearby.

The ubiquity of sound, and our frequent inability to shield ourselves from it, can lead many of us to take it for granted. We can easily shut out the visual world; if simply closing your eyes doesn't cut it, then covering them with your hand or a thick piece of cloth will do the trick. When it comes to sound though, it just has an in-your-face quality to it that is difficult for any person with normal hearing to disregard.

The fact that sound at least *appears* to be nearly inescapable has also figured into the Daredevil mythos in various ways. In the 2003 *Daredevil* movie, starring Ben Affleck, viewers were shown how the title character had to sleep in a sensory deprivation tank in order to escape the din of the city! While this idea wasn't taken from the comics, longtime readers are reminded in other ways of the sometimes oppressive role of sound in Matt's world.

There are storylines that deal with his senses going haywire, where sound tends to be the biggest concern. Mentions of young Matt's guru Stick also tend to serve as reminders of his role in helping Matt "control" his senses. We see an example of young Matt's struggle with sound in the *Daredevil* television show as well.[1] However, the show doesn't make the case that Matt's heightened hearing bothers him *as an adult,* or under normal circumstances, and neither do the earlier comics. The first indication that even sound at normal levels might spell trouble for Daredevil appeared as late as *Daredevil* #125, by Marv Wolfman and Bob Brown. A caption reads:

"*Night*: A time of supposed *quiet*, when sleep comes readily for most Manhattanites, but *not* for Matt Murdock. *Imagine* what agonizing *noises* his super-sensitive hearing must accidentally overhear. *Voices! Cars! All* of the mind-shattering *sounds* which most of us normally block out."

There is a real condition called *hyperacusis* which is defined as an increased sensitivity to sound. Ordinary noises may be perceived as abnormally loud, annoying, or painful. What hyperacusis in the real world *doesn't* do, however, is make a person objectively more sensitive to sound on a physical level, like we expect of a fictional character like Daredevil. It does seem plausible that hearing a wider range of sounds would predispose someone to being more bothered by noise, but I don't think we can assume that this would necessarily be painful.

The idea that our title character is able to tap into a very far-reaching and extraordinarily rich world of sound is something I associate more with the recent comics than the early ones. With the exception of egregious examples like Daredevil landing a spaceship in Central Park by listening to the absence of heartbeats in *Daredevil*

#2, the early comics didn't usually push Matt's sense of hearing very far. In *Daredevil* #7 (1964), we even see Daredevil use a "snoopscope" billy club attachment to pick up sounds around him. Where certain early abilities such as sensing colors by touch have faded away entirely (see chapter eight), Daredevil's sense of hearing seems to have become *more* extreme over the years.

As far as I can tell, it was Frank Miller who introduced the idea that Daredevil can sift through sounds of seemingly any intensity for several blocks to find just the one he is looking for. He does so in a classic scene in *Daredevil* #169, where Daredevil concentrates to listen for the sound of a single cough in order to find Bullseye and the people he's holding captive. Matt is thinking to himself:

"Hmmm... Bullseye smokes cigarettes, and he's hiding with someone who has a bad throat... A condition that would get worse without the lozenges. [...] Okay, DD, so you're a hot shot, super hero. So you can touch, taste, smell, and hear, better than anyone else on Earth. But can you detect a single *cough* in the ocean of noise below you?" Alongside panels of various sounds of the city, ranging from honking cars to the drip of a leaky faucet, the captions read:

"Daredevil relaxes, and clears his mind of thought. A wave of sound roars up from the street, strong and clear. He shuts it out. He concentrates. Softer sounds murmur to him from a thousand separate sources. He shuts them out. He *strains*. Still softer sounds whisper faintly. He sifts through them carefully, isolating each. Finally he hears it: He smiles."

Since this ability was introduced, Daredevil has done similar things on other occasions, and the Daredevil show also has Matt listen for faint sounds from impossibly far away. One scene that comes to mind can be found in *Daredevil* #105 (1998), by Ed Brubaker and Michael Lark, where we see Matt search for Mister Fear in ways that are reminiscent of Miller's earlier issue. In *Daredevil* #1 (1998), by Kevin Smith and Joe Quesada, Matt notices a pair of heartbeats(!) from blocks away.

I probably don't need to explain why these examples count as major "stimulus problems." However, I would point out that there are also narrative problems with this particular *deus ex machina* device that create an awkward contrast with how Daredevil otherwise operates. If he can do this, why isn't this his default mode of operating? Why is it the thing to go to when all else fails as opposed to the *first* thing you do? To be clear, I am not talking about Matt detecting a loud scream from some victim a few blocks away, but the nearly transcendental ability to tune into faint conversations inside distant buildings.

I think it is probably difficult to get away from *some* laws-of-physics-breaking when it comes to Daredevil's sense of hearing, and the realm of sound more broadly, but I firmly believe that it is a generally good idea for creators to be mindful of how this sense is used, for reasons of narrative consistency if nothing else.

I will get back to the physical limits of sound in a little bit, but let's first get a closer look at the human ear!

THE EAR – AN AUDIBLE CRASH COURSE

Hearing, like seeing, is a highly specialized perceptual phenomenon that depends on some remarkably sophisticated hardware. We have already briefly touched on some of the workings of the inner ear and the auditory pathway to the brain, but there is more to this story.

If we start from the outside and work our way in, we'll first note the outer ear, also known as the *pinna*. This vaguely funnel-shaped oddity may not look like much (unless we dress it up with some earrings), especially compared to the outer ear of most of our mammalian relatives. However, the pinna helps direct sound into the ear canal and is also hugely important to our ability to determine the location of sound in the vertical (up-down) plane for which we rely on so-called spectral cues. The strange folds of the pinna give the incoming sound a characteristic pattern depending on the angle at which it falls.[2] This means that if you're Matt Murdock, you want to make sure to not obstruct or alter the shape of the outer ear in any way. Daredevil costume designers beware!

Once the sound has entered the ear canal, it continues to be affected by the physical dimensions of the body, and the size and shape of this passageway to the middle ear actually help to amplify sounds of frequencies between 2,000 and 4,000 Hz. Since Daredevil looks much like a typical human specimen, we can safely assume that his ear canals will be doing something very similar. When the sound reaches the eardrum and sets it in motion, it has officially arrived at the middle ear at which point the vibrations, in turn, move the three tiny bones – the malleus, incus, and stapes – which connect the eardrum to the inner ear.

You may be wondering why we even have these bones, collectively known as the "auditory ossicles," in the first place. Why doesn't the sound pressure wave just travel directly into the inner ear? The short answer is something called *impedance matching*. And since this will come back to haunt us on the topic of hearing heartbeats, we might as well get into it. Because the inner ear is filled with fluid that doesn't compress as easily as air, the sound cannot easily cross the border between the different mediums. This is also why you can't hear much of what's happening above the surface when you are underwater, as most of the sound in the air is reflected. The ossicles in the middle ear help us overcome this difference in *acoustic impedance*, by working as a kind of lever and narrowly transferring the force from the outside to the oval window, the membrane that provides the barrier to the inner ear. Pretty clever, huh?

As we've addressed previously, the workings of the inner ear are contained within a small snail-like structure called the cochlea. If you were to take it out of the skull and straighten it out, you would find that it's divided lengthwise (almost all the way to the end) by a piece of tissue called the basilar membrane. Near the oval window, we find the round window which marks the end of the fluid-filled chamber which surrounds, and snakes around, the basilar membrane. When the stapes pushes on the oval window, this is matched by the movements of the round window. This helps the fluid move without having to be compressed more than necessary. (I like to picture this looking something like a plunger in a toilet bowl.) The movements created by the sounds that are transferred to the inner ear create

distortions of the basilar membrane at specific points that corre-
spond with the particular frequency.

We've already talked about how it is the hair cells that convert the
mechanical energy of sound into electrical signals, and that these
hair cells are arranged xylophone-style along the basilar membrane.
The sound isn't acting on the hair cells *directly*, it's more the case that
the physical distortion of the basilar membrane creates shearing
forces that activate the hair cells. The higher frequencies are detected
at the base of the cochlea, nearest to where the sound vibrations
enter, and the lower frequencies at the so-called apex.[3]

Hair cells are "maximally tuned" to specific frequencies, but they
still respond to vibrations above and particularly below their
preferred frequency. In fact, this is one way that the intensity of
sound can be encoded. The louder a sound is at a particular
frequency, the greater the activation of hair cells at that frequency *and*
of neighboring hair cells will be. This makes sense considering that a
more intense sound will cause a greater amount of distortion of the
basilar membrane.

There are also other ways for hair cells to communicate the intensity of the frequency they're responding to. Each of these hair cells is
connected to between ten and twenty "type I" nerve fibers which
carry the information from the ear onward. This may sound awfully
wasteful – even the cones in the retina are not that well supplied –
but it makes sense when you consider the limitations of the action
potential, as well as the short timescale on which sound operates. A
nerve fiber can only fire so many times per second, and the sound
event the particular hair cell is responding to can be very brief. If you
have more nerve fibers to carry the load, the intensity of the sound
can be relayed by the *proportion* of type I nerve fibers firing at any one
time, rather than the less trustworthy activity of a single nerve fiber.

The scenario above is true for frequencies above a few thousand
Herz. At lower frequencies, each wave of sound hits "its" spot of the
basilar membrane at rates that are low enough that the ion channels
of the hair cells can open and close with *each cycle* of the wave, and
both the hair cells and the auditory nerve fibers can fire in lock step

with the distortion of the basilar membrane, which is a phenomenon called phase locking. But if this firing pattern is reserved for signaling the frequency of the sound, not its intensity, how *do* you signal intensity? This conundrum is solved by having the nerves skip firing for many or most cycles of the sound at low intensities and skipping very few at high intensities. In this case too, it is helpful to have many nerve cells connected to each hair cell.

One final, and pretty darn exciting thing we need to cover about how the ear works is that there are actually *two* different types of hair cells. The ones I've talked about so far are actually called *inner* hair cells and they tend to get most of the attention, seeing as they're mainly responsible for actually registering the sound. However, there is also a class of hair cells called *outer* hair cells. These are only found in mammals and function as superstar cheerleaders. They actively help *amplify* the sound through a sort of feedback mechanism.

The outer hair cells make a big difference to the amount of sound we can hear, and crucial to their function is a special protein called *prestin*. One study found that the deletion of the gene that codes for prestin in mice amounted to more than a hundred-fold loss of hearing sensitivity, or the equivalent of 40-60 dB.[4]

Another interesting thing about prestin is that there are a handful of specific sequence variations in this protein that are found in mammals that happen to specialize in high-frequency hearing, such as certain whales, dolphins, and echolocating bats. While whales and dolphins are close relatives, neither one of them is closely related to any member of the bat family. This suggests that the similarities between the prestin gene in these animals have not arisen as a result of common ancestry, but have been driven by so-called parallel evolution to meet the demands of a particular lifestyle.

Obviously, the very fact that we have a candidate gene that is linked to hearing sensitivity generally, *and* can potentially unlock the key to hearing unusually high frequencies, is very exciting if you are looking to run a thought experiment on a certain fictional superhero.

· · ·

As amazing as the mammalian ear is, it is worth considering the complexity of the task it's up against. The real world is full of many different sounds of varying origins and qualities that are essentially superimposed on each other. Remember that the basilar membrane registers the totality of sound at any given moment, and there is (at this point) none of the separation into different "objects" we associate more readily with the visual world. An image projected onto the retina contains the physical separation of things in two-dimensional space. The auditory system works nothing like this. To quote Jan Schnupp, Israel Nelken, and Andrew King, authors of *Auditory Neuroscience: Making Sense of Sound*:

"To understand the problem that this superposition of sounds poses, consider that the process that generates the acoustic mixture is crucially different from the process that generates a visual mixture of objects. In vision, things that are in front occlude those that are behind them. This means that occluded background objects are only partly visible and need to 'completed'; that is, their hidden bits must be inferred, but the visual images of objects rarely mix. In contrast, the additive mixing of sound waves is much more akin to the superposition of transparent layers."

One reason I wanted to spend the time on the minutiae of how the cochlea registers different aspects of sound is that it helps us get a sense of what kind of information the rest of the auditory system has available to it. Despite the fact that the ear itself has no way to distinguish one sound from another, later processing allows us to, for instance, pick out and follow a single voice in a crowd, or attend to any other sound we might be interested in. We typically experience different sounds as distinct from one another, which is an amazing feat, especially since a particular sound event typically consists of several different frequencies.

The phenomenon that allows us to pick out a single voice in a crowd, or the sound of a single cough in Hell's Kitchen (ahem), is called the cocktail party effect, after a 1953 paper by the cognitive scientist Colin Cherry. This phenomenon can be seen as a subtask of so-called auditory scene analysis, the process by which sound is segregated into individual streams.

One feature of the auditory system that is helpful in this regard is one we've talked about already: The fact that we have one ear on each side of the head and an auditory system that is able to use the difference in information between the ears to locate sounds in three-dimensional space. As mentioned previously, a single ear can also perform some functions of spatial hearing, but not to the same degree.

Using spatial hearing would obviously be a good way for the brain to figure out which features of the soundscape belong together. And this is certainly a contributing factor. However, there are other cues as well. One is the difference in fundamental frequency between two sounds such that two sounds are easier to separate if the difference between them is large.[5] Another factor is the *temporal* aspect; a group of frequencies that *begin* at the same time can be interpreted as belonging together. Ultimately, there is probably even more going on here. Schnupp, Nelken, and King again remind us of the difficult problem the brain is up against:

"Mathematically, [decomposing a sound waveform into the sum of waveforms emitted by multiple sources] is akin to trying to solve equations where we have only two knowns (the vibration of each eardrum) to determine an a priori unknown, and possibly quite large, number of unknowns (the vibrations of each sound source.) There is no unique solution for such a problem. The problem of auditory scene analysis can be tackled only with the help of additional assumptions about the likely properties of sounds emitted by sound sources in the real world."

If we bring all of this back to Daredevil, we might ask what could be done to bestow him with a heightened sense of hearing, if we want to take inspiration from real-world biology. I have already mentioned prestin, that crucial protein found in the outer hair cells. We would definitely want to give Matt the mutated version that might give him better high-frequency hearing. And perhaps giving him more than the normal number of outer hair cells might help things along too.

You may also recall the findings from chapter three, showing that cats have an unusually impressive hearing range and that the number of turns of the cochlea has been associated with better than other-wise expected low-frequency hearing in mammals. Let's make Matt the human version of a cat! We could also play around with the length and other characteristics of the basilar membrane itself and the number of *inner* hair cells, in an effort to improve the frequency resolution.

What if we additionally gave Matt even more nerve fibers connected to each of these inner hair cells? This might theoretically improve the resolution of sound *intensity*. Which other aspects of the performance of the inner ear would we want to play with if we could? Because time is such a central concept to sound and hearing, it would be useful for the ear to know exactly when a sound begins and when it ends. I am, of course, talking about *temporal* resolution.[6]

Every way we could think of to improve the quantity and quality of Daredevil's sense of hearing would make him better at all of the subtasks that hearing permits, including separating different sound streams from each other and detecting echoes and other faint sounds. And the miracle exception means we wouldn't even have to explain all of it! However, the biggest obstacle to many of the examples of super-hearing you see in the comics isn't the ear, but – again – the nature of the stimulus.

THE THINGS WE CAN(NOT) HEAR

It would be much easier to bestow Daredevil with *insanely* good hearing if it weren't for two small problems. The first, that we have hinted at already, is that sound doesn't travel infinite distances or through an unlimited number of obstructions. The second is that even a perfectly average human hears *really* well. At least across those frequencies that we are most sensitive to. For frequencies between 250 and 8,000 Hz, a young person with normal hearing can hear sounds as soft as 10 dB. For frequencies around 2,000-4,000 Hz, that same person can hear sounds below 0 dB. Remember, this is the frequency range that is also amplified by the ear canal.

You may be wondering how it is that anyone can hear below zero *anything*, but this is where we need to keep in mind that a lot of the units we humans use to measure things tend to not only be anthropocentric, but often arbitrary as well. "0 dB" does not indicate some absolute zero of sound, but is set to coincide with the typical human hearing threshold at 1,000 Hz specifically, and standardized to coincide with a pressure deviation of 20 μPa (micropascals). The pascal is a unit of pressure and tells us about the *deviation* from the average ambient air pressure that a particular sound pressure measurement represents. For a longitudinal pressure wave in a medium like air, these deviations in sound pressure are also a measure of the amplitude ("size") of the wave.

The number given in pascals is a measure of something out in the real world, while the decibel (dB) is completely relative. It is so relative in fact, that I really should have been using "dB (SPL)," where SPL stands for sound pressure level, throughout this book to indicate that I was talking about sound specifically because the dB can be attached to anything. Regardless of what it measures, many of us are scared of the decibel because we associate it with some really weird math. The decibel scale is logarithmic, not linear, and this makes using it kind of tricky. If we are talking about sound *pressure*, each step of twenty along the scale represents one order of magnitude, so that a sound at 40 dB has a sound pressure that is ten times greater than one at 20 dB, and one hundred times greater than a sound at 0 dB.[7]

If you think that this is just an evil plot to torture high school students, you will be happy to know that there is actually more to it than that. It seems that our subjective sense of the *loudness* of sound tracks more closely with a logarithmic scale than a more conventional linear scale. For instance, a typical 60 dB conversation certainly doesn't *sound* more than thirty times louder than a whisper at 30 dB. The inner ear itself doesn't represent sound linearly either, which is why I compared the hearing ranges of different animals using the (similarly non-linear) octave, back in chapter three. Using a loga-rithmic scale also saves us from dealing with a ridiculous number of zeroes, so it at least swaps one form of torture for another.

Despite the math, the dB scale for sound level is the one that most people are familiar with, which is why I've been pretty casual about using it so far. Most people have seen the little charts of typical sound levels – 90 dB is bad, 120 dB is *painful*, 40 dB is the sound of a quiet library, and 20 dB is the rustle of leaves, and so on. It's relatable, and that's why I'm going to continue using it.

The first question we may want to answer is: If 0 dB is not the "absolute zero" of sound, what is? Is there such a limit? There is. For a sound pressure wave to meaningfully exist (whether we can hear it or not), there needs to be a deviation in localized air pressure above "background levels." The background is set by the random motion of air molecules (that move in a pattern called *Brownian motion*). As a sound becomes fainter, from absorption by objects or the air itself, and from how it is spread ever "thinner" when it travels, the pressure wave will lose cohesion and the tiny movements become smaller until it disappears into the background, like the waves from a pebble thrown into a pond. Under typical conditions, this "absolute zero of sound," comes out to the equivalent of -23 dB.

With all the logarithmic math going on, this may sound like it's a lot lower than the lowest human hearing threshold (which is as low as -9 dB at 3,000 Hz), but using fancy math – or better yet, an online conversion tool – we note that this amounts to a pressure deviation that is only five times greater than the random oscillations of air molecules. Our sense of hearing comes awfully close to the limit of what is theoretically possible. This is great for the average person but doesn't buy a whole lot of extra room for Matt Murdock's super-hear-ing. However, while it may not get us a lot of extra mileage at the human sweet spot of hearing, there may be more room to play at lower and higher frequencies.

But if we're going to translate this bit of insight into answering questions like "How much farther away can Daredevil be from a sound source and still hear it?," we're going to have to do just a *tiny* bit more math, and look at how the sound level drops the further away you get from a sound source. For practical purposes, we are going to assume that the sound source is radiating in all directions from some surface. This means that the sound *pressure* will drop by half when you double the distance from the source.[8]

But how do we plot this kind of thing on a decibel scale, with its weird non-linearity? Well, if you are standing three feet (or one meter) away from a friend having a conversation that registers to your ears as 60 dB, and you double the distance between you to six feet, the sound pressure level at your ears will be 54 dB! For mathematical reasons I won't bore you with, each doubling of the distance from the sound source in question translates into a drop in sound pressure of around 6 dB, or 6.0206 if you want to be precise.

Suddenly, things are starting to look up! Going by this calculation, the difference between -9 dB (best human threshold at 3,000 Hz), and - 23 dB (theoretical limit), means that someone like Matt Murdock could be more than four times farther away than the average person and still hear the same sound, provided that we push his hearing threshold to the absolute max! For other frequencies, such as well below 250 Hz, and well above 8,000, where normal human hearing is much worse, there is even more room for improvement. *But*, there are a couple of caveats to consider.

The first problem is that the farther away you are from a sound source, the more likely it is that the sound will encounter something on the way that absorbs it, scatters it, knocks it off course, or does any number of things that will prevent it from reaching you. The second problem is that one of the obstacles the sound has to deal with is the *air itself*. This is not really a problem for low-frequency sounds which, as you're well aware at this point, will move relatively easily through walls or past obstacles. There's some loss of energy, and some of the sound is reflected, but this is much less of an issue than for higher frequencies

For high-frequency sounds, the process can be brutal. For instance, a sound at 20,000 Hz loses around 45 dB over a distance of 100 m (or roughly 325 feet) from so-called attenuation *alone*. This has a very big impact on critters such as echolocating bats. They are indeed marvels of nature, but the range of the sounds they emit is very short. Add to this that they also need the sound to be loud enough so that the faint echo can make it back, and we're talking about a range of at most a few meters for the detection of insects. Other kinds of high-frequency animal communication, such as that of mice, is also a rather intimate affair.[9]

A final thing that some knowledge of "decibel math" helps us put into perspective is how fainter sounds fare against louder sounds. If we look again at the sound of a single cough – or the sound of a leaky faucet – at some relatively remote distance, we are not only dealing with a sound that, on its own, may not meaningfully exist that far away from its source. There's also the issue that even if it does, we have the added complication of its having to compete with other, much louder sounds. If we do the math here and remember the fact that all sounds are presented to the basilar membrane *superimposed on one another*, it is easy to see why very faint sounds become difficult to detect.

For instance, if we pool together two sounds of a particular frequency, but at very different intensities, we cannot just add them together. A 20 dB sound added to a 40 dB sound will *not* result in a 60 dB sound. The numbers here are not linear, and the fainter sound is in fact *much* fainter than the louder sound. I will spare you the math, but the number you get when you add these two sounds, using the numbers hiding behind it all, is actually 40.8 dB! That's not a big difference compared to 40 dB.

Sadly, this insight also means that being able to hear down to that theoretical -23 dB won't actually give you much more useful hearing since no natural environment is even *nearly* that quiet, and the very softest sounds would be lost in the background noise anyway.[10]

WHAT DAREDEVIL ACTUALLY HEARS (MAYBE)

My spidey sense is telling me that you're about to throw this book out the window. Sure, our typical human ears are great and all that, but is there *nothing* that a little super-hearing could do for you? Some of you may even be thinking that I'm coming down too hard on the impossibility of hearing things from very far away. What about those remote sound-detecting devices that can record conversations from inside a room from blocks away? Isn't *that* proof that Daredevil should be able to do the same?

Again, I'm sad to disappoint you. The more extreme spy-level devices you may have seen only actually measure sound *indirectly*. By using laser technology, they can detect the vibrations of a surface induced by sound locally. This is all very cool but also very much not part of Daredevil's power set. But all is not doom and gloom. While I do want to get across that much of what Matt Murdock supposedly hears in the comics is so far beyond impossible that we might as well be calling it magic, I absolutely do *not* want to take away anyone's sense of wonder. And that includes my own.

The next chapter will look at the specialized hearing task that is human echolocation, for which the ability to hear fainter sounds across a wider range and with higher resolution would obviously be useful. But first, let us close out this chapter by considering what kinds of sound sources heightened hearing above and below the human sweet spot would give a person access to. And here, I will invoke the miracle exception so that we don't have to be bound by sensory biology to the same extent as the physics of the stimulus.

Let's start with the higher frequencies. Since these sounds don't travel very far, I would seriously question the usefulness to Daredevil of being able to hear sounds at frequencies much higher than one octave above normal human hearing (that is 35-40,000 Hz). Since Matt Murdock, unlike a bat, also doesn't need to go hunting his own insects for dinner it makes more sense to focus on the kinds of sounds that arise from human activity and from at least some distance away. If we accept that we need to be focusing on things relatively close by, there are still obvious advantages to hearing higher frequency sounds. The natural world of winds blowing through trees, and critters chirping is of course a rich source of sounds across a wide spectrum of frequencies. So are other naturally "noisy" processes.

While the everyday use of the term noise is often reserved for unwanted or intrusive sounds, this isn't necessarily the strict physical definition. In physics, noise typically refers to a random mix of different frequencies with a particular spectral pattern and is not limited to descriptions of sound.[11] You may be familiar with white noise, which is a combination of equally intense frequencies of sound across the spectrum. Pink noise also contains a mix of frequencies, but with a relative emphasis on low-frequency content because there is a steady decline in sound intensity with an increase in frequency. There are other "colors" of noise as well, but for many natural phenomena, pink noise is a good one to look at (and also a great name for rock a band). Among the natural sounds that have this noise pattern are waves crashing on the beach, the rustle of leaves, and rain. These are all sounds that the average human can also hear, but it's interesting to consider how they might sound to Matt Murdock and to point out that many sounds extend beyond the human range.

Since most noise-control efforts (I'm now talking about noise in the more casual manner) tend to focus on the human hearing range, we also find many household products that generate high-frequency sounds that will go unnoticed by the human occupants of a home but might be disturbing to cats, or dogs. Or Matt Murdock. *CNET Magazine* ran an article in their Spring 2018 edition for which they had looked for ultrasound (sounds above 20,000 Hz) in a typical media room. They found that a turned-on LED TV – with nothing playing on it! – and an LED light source both added significantly to the ultrasonic background noise of the room, which even at baseline had a pretty significant band of background noise between 20,000 and 30,000 Hertz.[12]

By the same reasoning, it seems plausible that Daredevil may be able to hear other kinds of electronic devices, such as alarm and surveillance systems. In *Daredevil* #78 (1998), by Brian Michael Bendis and Alex Maleev, Matt claims confidently that he can detect "recording devices" (and that he is able to do so with his "sensory radar"), which doesn't make much sense as a blanket statement. The device in question would first have to make a sound he can hear and also be exposed enough so that the sound can actually reach him. And so, it does make sense in theory, provided that certain conditions are met.

This takes us to another thing we've seen Matt do, which is to follow the sound of electronic circuitry, such as in the scene which takes him and Elektra to a hidden door in "*Regrets Only*," the sixth episode of the second season of the *Daredevil* television show. Several factors contribute to the creation of audible noise in electronic equipment (including wiring), and you might be familiar with the "hum" of power lines and the buzz of fluorescent lighting. These sounds are obviously not limited to higher frequencies but make for another source of information of value to anyone with a heightened sense of hearing, especially if they happen to be blind.

The sound sources that Matt Murdock should be almost literally bathing in, though, are of the lower frequency variety, especially since this is another part of the spectrum where the average person's hearing is relatively poor. Imagine the rumble of the New York subway system, street traffic, and heavy machinery. On the one hand, this would potentially be quite annoying. On the other hand, this too would be a great source of information that could serve as easily audible environmental markers.

Then there is of course that other famous Daredevil sound, that of the beating human heart! Like other sounds that are generated by the human body, such as breath sounds and those embarrassing rumblings of the gastrointestinal system, heartbeats generate most of their audible output in the lower frequencies. In fact, only a small sliver of the sounds generated by the heart is audible even to a physician with a stethoscope.[13] Heart sounds range from below 20 Hz to around 500 Hz, but most of the energy is focused in the lower range, and the loudest of the two primary heart sounds (the so-called first heart sound) is strongest around and below 100 Hz, where the typical human has a very high hearing threshold.

So how "loud" is this sound? Well before addressing that question we need to remember the complication of acoustic impedance. When a sound passes from a denser object (a human body) to the air, or the other way around, this limits the fraction of the sound energy that is actually passed from one medium to another. As alluded to above, there are obviously sounds generated inside the body that we can in fact hear with our standard-issue hearing (and we can amuse ourselves by thinking about how Daredevil might pick out a hungry ninja). But there is still a difference between listening to a heart with a stethoscope and listening to it from a distance that goes beyond the fact the former is closer to the sound source.

And here's where I'm going to let you in on a secret: The amount of research done to measure the sound output of the human heart, as measured *outside and away from the body* with equipment that picks up sounds that *no real human can hear*, is pretty close to zero. This topic may be of huge interest to people like yours truly, but obviously isn't something of major interest to medical science.[14]

But all is not lost. We do know that the impedance factor is less of a problem with those low-frequency sounds that Matt would be relying on to hear the beat of the human heart. We also know that the tiny vibrations produced by the heart *at* the chest wall would likely fare better than the heart sounds that come about as a result of blood flow and movements of tissues deep inside the body. The latter have to move *through* the body whereas the former would stem from the outside of the chest vibrating the air.

To put a cap to a potentially even longer story, it does seem plausible that the sound of the human heart would be audible outside the human body to someone with exceptionally good low-frequency hearing, especially at the lowest frequencies. However, this is still a relatively soft sound and one that would have to compete with other sounds in the environment. Hearing a heartbeat in the same room, a few feet away is one thing, hearing it blocks away is quite another.

In the *Daredevil* comics, hearing heartbeats has two main uses. The first is to recognize people, and the second is to detect whether or not someone is lying. While the heartbeat could theoretically work as a biomarker, at least according to the Pentagon who are looking into using remote heartbeat detection to identify people (with an infrared laser, *not* a microphone), it has always seemed odd to me that Matt Murdock wouldn't use other ways to recognize people.[15] Everything from voice, to gait, to scent would seem more obvious than the detection of heartbeats!

When it comes to using heartbeats to detect lies, there is instead the caveat that even regular polygraph tests don't actually detect lies *per se*. They detect subtle behaviors and bodily responses *associated* with lying. This means that someone can theoretically appear to be lying even when they aren't, and vice versa and the polygraph as a tool has been criticized for these shortcomings.[16] As an interesting aside, there is a connection between the classic lie detector and the history of comics. The psychologist William Moulton Marston (1893-1947) was the inventor, with his wife Elizabeth Holloway, of an early prototype of the lie detector. He was also the creator of Wonder Woman!

Regardless of the frequency, or mix of frequencies, that we are talking about, it seems to me that the most useful aspect of being able to hear a wider range of softer sounds would be that it makes active sound sources of many of the objects that we consider to be otherwise silent. The steady hum of the refrigerator would be matched by the sound of fluorescent lighting or (to us) inaudible sounds of home electronics or the low rumble of a neighbor dragging a sofa chair across the floor. A world with more sound sources in it that can tell Matt Murdock what some of these objects are, and where, would create a richer tapestry of sound that could be used to recognize a familiar place, or figure things out about a space and the identity of various objects in it.

It is probably inevitable that you leave this chapter less confident in Daredevil's ability to do things with his ears that lie very far beyond human norms, even when there would be very interesting uses for a slightly less extreme take on this sense. I get the appeal of having Matt eavesdrop on ordinary conversations on the other side of a thick wall, as impossible as this is. Because of how unforgiving the physics of sound and hearing can be, I'm less uncomfortable with minor "stimulus problems" in the realm of sound than I am in most other Daredevil-related domains. However, the extremes of hearing we sometimes see are not usually necessary to the plot, and many such scenes could benefit from some balance. Imagining what it's like to hear more is exciting, whereas imagining that sounds and ears are limitless is less so.

Next, we will look at what makes it possible to not only hear more active sources of sound but objects that are otherwise silent. This is an area of research that is interesting enough when applied to the real world, so you can imagine what might happen when we add "super-hearing" to the mix.

CHAPTER 6
A SENSE OF SPACE

 "The blind man of Puisaux judges of his nearness to the fire by the heat, and of a vessel being full by the noise made when pouring in a liquid; and he judges of his nearness to objects by the action of the air on his face. He is so sensitive to the least changes in the currents of air that he can distinguish between a street and a closed alley."

A Letter About the Blind for the Use of Those Who Can See, by Denis Diderot (1749)

Since the time of Denis Diderot, the mysterious ability of at least some blind people to sense the presence of objects around them has been the subject of intellectual curiosity. However, until the middle of the twentieth century, no one was able to definitively pin down what exactly gave rise to the ability. There were many different theories, ranging from a heightened sensitivity to pressure, or temperature, of the skin on the face to the development of a *bona fide* "sixth sense."[1] Diderot himself clearly assumed that it was the "action of the air on his face" that gave the blind man of Puisaux the ability to detect objects.

We now know that "facial vision" depends on the sense of hearing. That is not to say that people are unable to detect changes in temperature when moving in relation to a heat source, or that we cannot feel the draft coming from the side when walking past an open door, or the corner of a building where the wind, kept at bay by the nearby wall, can suddenly be felt blowing in over an open street. These are obviously real sensations that contribute to a complete sense of our surroundings, and would naturally be useful to people who are blind, including our own Matt Murdock. However, these sensations are not central to the more narrowly defined "sense of obstacles" in the absence of vision.

THE CORNELL EXPERIMENTS

The man who led the research to settle the matter of facial vision once and for all was experimental psychologist Karl M Dallenbach, then at Cornell University. Together with his graduate students, he conducted a series of in-depth experiments that each added a piece of the puzzle.[2] As luck would have it, their first study, conducted in 1940, was actually filmed and can (and should!) be viewed online.[3] The footage doesn't add much in terms of allowing us to understand the experiments, which are described in great detail in the final paper, but it does contain an opening scene of Dallenbach and his two graduate student collaborators Michael Supa and Milton Cotzin just standing around having a smoke! Why such a seemingly random scene would make it into the official record of this groundbreaking experiment is a mystery, but a rather amusing one, especially to modern audiences.

Michael Supa would find himself in the position of both collaborator and subject for the study, as he had been blind since infancy, and made use of what was still termed "facial vision" in his daily life. The other three subjects included an undergraduate student named Edward Smallwood, who was also blind and highly proficient at detecting obstacles, as well as two sighted graduate students, one of them Dallenbach's own son. Going into the experiment, the participants described their own theories and expectations:

"The blind [subjects] possessed the ability to perceive obstacles from a distance and utilized it to a marked degree in their daily lives. Neither, however, could explain the basis of his judgment. [Michael Supa] thought that audition helped, but [Edward Smallwood] on the contrary was of the opinion that sounds hindered. The two sighted [subjects] were unable, at the beginning of the study, to detect the presence of obstacles when blindfolded; and they expressed grave doubt concerning their ability to learn to do so – but both were willing to try."

They were first subjected to an exploratory round of experiments in which they were asked to indicate when they first became aware of the *presence* of a wall (experiment one), or a masonite board placed in the same room (experiments two and three), and when they were as close to the obstacle as they could get *without actually touching it*. These two distances from the obstacles were referred to as "first perception" and "final appraisal." For each attempt, the subjects would be led to starting points at varying distances from the final target, and the experiments also consisted of different series where they either wore shoes or walked in only their socks. Each trial would last until the subjects had either made their final appraisal twenty-five times without touching the target or collided with it fifty times.

Not surprisingly, the sighted subjects initially failed to impress. During the first experiment, they didn't notice the wall they were approaching until they were almost about to walk into it and actually did walk into it several times. The blind subjects, on the other hand, were much more successful, and were also seemingly immune to the disorientation efforts of the experimenter:

"They did not run into the wall a single time in 25 successive trials, nor did they require guidance in approaching the wall. Stepping out unhesitatingly, they walked forward in a straight line. As they both reported, their perception of the side walls enabled them to follow the path to the end wall that was equally distant from the sides."

However, during the course of these preliminary experiments, two things became very clear. The first finding was that the sighted subjects gradually acquired the skills necessary, and proved to be very fast learners. So fast, in fact, that the effect of practice outweighed the objectively greater difficulty of the later experiments. The effects of learning were evident even during the course of the first experiment. When it came time to take their shoes off, the blind subjects did worse than they had with them on. The sighted subjects, on the other hand, actually did better, while still trailing the blind subjects in overall performance.

The other key finding was that, after accounting for the sighted subjects' impressive learning curve, approaching the obstacle with socked feet adversely affected performance. That meant that the study, right from the outset, yielded strong indications that facial vision depended on sound to a significant degree since removing the shoes made the sound of the footsteps softer. This suspicion was confirmed in the second round of experiments when the subjects attempted to approach the masonite board with either their heads covered by a thick veil, their ears plugged shut, or their useful hearing effectively masked by a tone in their ears.

The veil, which was not in touch with the subjects' skin, but suspended by something resembling the brim of a hat placed on their heads, did negatively affect performance, though it was believed this was due to its effect on the incoming sound. Removing all sound, however, completely destroyed the ability of all subjects to get any sense of the location of the board, which they all collided with in every single attempt. The same thing happened in the so-called sound screen experiment, in which a continuous, moderately intense tone of 1,000 Hz was played in headphones worn by the subjects.

In a final experiment, the subjects were placed in a separate soundproof room where they listened to a live feed of experimenter Milton Cotzin approaching the board with a microphone. This time, they were able to locate it once again, based only on the incoming sound of *someone else's* footsteps and the reflected sound.

This groundbreaking study would end up being featured in a 1941 *Life Magazine* article with the definitive-sounding title "'Sixth Sense' of Blind Discovered to be Hearing." Did this mean that the case was closed? It would appear so. But, there was still *one* further possibility to examine. What if *sound* was necessary, but the ability to hear it was not? Could it be, as some had previously suggested, that the phenomenon arose from the sensation of pressure against the eardrum, more so than the perception of the sound by the brain?

Dallenbach had decisively proven that the ear was the crucial organ, but ever the thorough experimentalist, he followed up his initial findings with additional experiments. In a 1947 paper, detailing a study involving deaf-blind subjects, he and his new graduate student collaborator Philip Worchel hoped to determine whether the *eardrum* might be able to signal the presence of objects in the absence of the ability of the brain to hear sound.[4]

However, none of the subjects, although selected for their impressive ability to travel independently, were capable of repeating the performances of the first subjects. Whatever resourcefulness this group of people exhibited in their daily lives, it soon became evident that none of them actually had the ability to sense obstacles. Nor were they able to learn it during the course of the experiment.

Interestingly, the one subject who had initially insisted that he did have an obstacle sense and that it manifested as a feeling of pressure on the face, was found to actually be using the residual vision he had in one eye as opposed to some other mysterious ability. That the information carried by one sense can be interpreted by the receiver as coming from another is an interesting phenomenon in itself, and a good indication of why it is that what we *now* know to be echolocation has been so very difficult to pin down historically.

EDWARD SMALLWOOD AND THE SUPER ECHOLOCATORS

For me, perhaps the most interesting detail from the experiments at Cornell, and certainly from the perspective of trying to make sense of Daredevil, is the case of Edward Smallwood who participated in the first study. Smallwood insisted, quite vehemently, that his ability to detect objects was *not* based on sound, and that sounds were actually a hindrance. Only in the face of overwhelming evidence to the contrary did he change his mind.

This also means that Smallwood, while obviously relying on sound without being aware of it, didn't actively attempt to make any of his own sounds, such as by clicking his tongue or snapping his fingers, in the course of his daily life. Obviously, his own footsteps would have provided guidance, but it also seems reasonable to assume that he must have been well tuned in to the ambient sounds around him as well.

What makes Smallwood even more interesting though, is how exceptionally good he was at detecting the obstacles in the experiment. He not only vastly outperformed the sighted subjects, but he also crushed his blind "opponent" Michael Supa, making the difference in performance between the two much larger than the difference between Supa and the sighted subjects. This was particularly evident in terms of the measure of "first perception," i.e. the distance at which the obstacle could first be detected. In fact, the experimenters initially seemed to have quite a bit of trouble even finding a distance from the obstacle where Smallwood couldn't immediately detect it:

"[Edward Smallwood] possessed a very keen 'sense of obstacles.' The average distance of his 'first perceptions' is 18.04 +/- 6.69 ft. This value does not, however, represent his true ability. Both the average and [mean variance] are artifacts of the experimental conditions. He always perceived the wall immediately upon being led to the starting-points at 6, 12, and 18 ft, and in most trials at 24 ft. [...] When his results were computed from trials starting at points beyond 24 ft., i.e. at 30 and 36 ft., the average of his 'first perceptions' was 25.62 +/- 3.56

ft. These values represent a truer picture of his ability to perceive obstacles at a distance than those given in Table I."

Smallwood's ability to detect a wall at the end of a hallway from a distance of around twenty-six feet (or just under eight meters) can be compared to Supa's ability to do the same, which hovered at around six feet (or just under two meters). This amounts to a stunning four-fold difference. While Smallwood's ability to detect the wall diminished when he approached it on socked feet – quite the opposite of what he himself had predicted – he was still able to do so at an impressive distance of just under eighteen feet on average, compared to Supa's three to four feet.

This striking difference in performance persisted throughout the study, even when the wall was replaced with the relatively smaller masonite board, and the various starting positions were randomized. For anyone concerned that Smallwood might be reporting false positives, there were tests for that too. None of the subjects reported detecting the presence of an obstacle when there was none.

We can safely assume that Edward Smallwood did not possess any heightened senses or superpowers, although in an interesting twist that should delight Daredevil fans, he did actually go on to become a lawyer, and even practiced law in New York City for a time.

What exactly distinguished Smallwood from Supa, whose performance would be matched in short order by the two sighted rookies, is impossible for us to know, though if you will allow me to speculate it may be as simple as the two men having happened upon different strategies during their acquisition of this skill, in terms of which aspects of the echo sound their brains are most attentive to. Perhaps Edward Smallwood's brain had stumbled upon some less obvious, but ultimately more useful, feature of how the sound changes when reflected off an object. Or perhaps his brain was paying attention to the exact same things as Supa's brain, but was able to process the signal more efficiently and reliably?

· · ·

Exceptional skill at echolocation has been described elsewhere. Samuel Perkins Hayes, who collected many of the theories surrounding "facial vision," in the decade before the Cornell experiments, mentions the case of Martin, another unusually skilled blind New Yorker, as reported by the physician who made his acquaintance:

 "Martin was a native of New York City and had been blind nine years. He was of a fearless and impetuous disposition, and went about over the city without a guide. He passed up, down, and across great thoroughfares frequently and only a few times collided with a bicycle, which vehicle he detested. I was with him on occasions when I marveled at the perfect freedom with which he walked along crowded streets, showing not the slightest timidity, and requiring no aid whatever from me...

I was amazed to see him cross Broadway at 14th Street with perfect ease, and imagine my astonishment when he shied around some timbers that had been set up across the sidewalk to prop the wall of a building undergoing repairs. He got on and off street cars without a blunder and made his way across narrow streets without betraying his blindness. He used no cane nor did he feel his way with his hands. Had I not known that he was actually blind I would have thought that he was feigning.

I asked him how he knew his way and avoided collisions, and he invariably told me that he did not know. He seemed to be guided by what I shall term a miraculous instinct super induced by a subconscious mental condition. [...]"

Another account that Hayes provides in the same volume, is by

W. Hanks Levy, himself blind, who describes his ability to detect obstacles:

> "Whether within a house or in the open air, whether walking or standing still, I can tell, although quite blind, when I am opposite an object, and can perceive whether it be tall or short, slender or bulky. I can also detect whether it be a solitary object or a continuous fence; whether it be a closed fence or composed of open rails; and often whether it be a wooden fence, a brick or stone wall, or a thick-set hedge. I cannot usually perceive objects if much lower than my shoulder, but sometimes very low objects can be detected... When passing along a street I can distinguish shops from private houses, and even point out the doors and the windows, etc..., and this whether the doors be shut or open. When a window consists of one entire sheet of glass, it is more difficult to discover than one composed of a number of small panes... When the lower part of a fence is brickwork, and the upper part rails, the fact can be detected, and the line where the two meet easily perceived. Irregularities in height and projections and indentations in walls, can also be discovered."

The Indian-born American author Ved Metha, who lost his sight at the age of four from meningitis, has provided another first-person account of his experience learning to echolocate when arriving at the Arkansas School for the Blind. It was something he took to with great ease:

> "One day in early spring, all the totally blind students were herded into a gymnasium and asked to run through an obstacle course. Plastic and wooden slabs of all sizes hung as low as the waist; others barely came down to the forehead. These slabs were rotated at

varying speeds, and the blind were asked to walk through the labyrinth at as great a speed as possible without bumping into the obstacles. The purpose of keeping the slabs moving was to prevent the students from getting accustomed to their position and to force them to strain every perceptual ability to sense the presence of obstacles against the skin - a pressure felt by a myriad of pores above, below, and next to our ears. Some of the slabs were of an even fainter mass than the slimmest solitary lamppost on a street corner. This obstacle course helped gauge how well an individual could distinguish one shadow-mass from another and, having located the one closest to him, circumvent it without running into yet another... The gymnasium was kept so quiet that the blind people could hear obstacles, although I could not help feeling that I could have run through the labyrinth with a jet buzzing over-head... For me, going through this obstacle course was child's play."[5]

It is always difficult to judge the veracity of some of these reports. Particularly in the case of Martin, whose skills were described by an outside observer, we have no way of knowing whether he was actu-ally completely blind, or how much of the story (if any) is simple hyperbole.

Without knowing the cause of his blindness, we also cannot rule out that it might have stemmed from an injury to the visual parts of the brain, rather than the eyes. A preserved pathway between healthy eyes and other parts of the brain can lead some people to develop a type of "blindsight" which helps steer a person around obstacles that are never perceived by the conscious mind. On the other hand, such unconscious vision is quite rudimentary and probably could not fully explain Martin's abilities. The point, though, is that we are lacking key information about his case.

What seems clear is that there are blind people who possess abilities that seem to be fundamentally incompatible with most commonly held expectations of blindness. Basic echolocation abilities can be attained by most people with intact hearing in a relatively short amount of time, whether they are blind or sighted, but there appear to be some who have been able to take their abilities to the extreme. The best way to think about Edward Smallwood, Ved Metha, and others like them, is perhaps as perceptual outliers, representatives of the theoretical limits of human capacity who may have simply stumbled upon exactly those cues which would provide the best sense of space, and fine-tuned their skills to match.

YOUR BRAIN ON ECHOES

From research conducted during the decades that have passed since the Cornell experiments, we've come to learn a great deal more about what actually underlies the human capacity to use echoes to detect silent objects and get a general sense of space. We now know that virtually all people with normal hearing *can* learn to echolocate and that it is a skill that can be picked up surprisingly quickly even though it obviously takes considerable time and effort to become highly skilled at it. Early blind echolocators may additionally be at an advantage here for reasons having to do with both neuroplasticity and the amount of time spent practicing.

Recent research has looked at which traits might influence the learning of this skill in sighted people new to echolocation. One study found that people who scored high on a test of "vividness of visual imagery" performed significantly better at the echolocation test than those with lower scores.[6] Another study, also involving sighted "echo-naïve" subjects, tested whether their ability to echolocate was associated with working memory, spatial ability, and skill at "sustained and divided attention." The latter, but not the former two were found to predict echolocation performance.[7] An additional factor that appears to be crucial is the ability to hear equally well in both ears, which ties in neatly with what happened to Matt in season three of the *Daredevil* television show when he temporarily lost his hearing in one ear, and with it his ability to detect objects.

Even without any training at all, we are naturally able to distinguish sound sources and reflected sounds, and a recent study looked at how this plays out in the brain.[8] The experimenters recorded the sounds of three sound events – a hand pat, a pole tap, and a ball bounce – in an anechoic environment (that is, one free of echoes). They then generated additional variations by combining these sounds with the reverberant sounds recorded in three rooms of different sizes – a kitchen, a hallway, and a gym respectively. They found that when the subjects correctly differentiated between sound *sources* the spike in brain activity would come sooner than when they were listening for, and correctly responding to, differences in room size. Reverberant information naturally reaches the ear a little later, and the subjects' brains knew how two differentiate the two.

Despite the fact that people can make the kinds of distinctions mentioned above, it is worth pointing out that our brains actively try to reduce echoes. Why would this be? Well, being served up a slightly delayed yet distorted kind of "after image" of the first sound interferes with the precise processing of that first sound and makes spatial judgments about the sound source more difficult. Our brains, trying to help us along, cleverly gives more weight to the first version of the sound through something called the precedence effect. This bias in favor of the first sound is reduced in situations where people are actively echolocating.[9]

As noted previously, far from all blind people *spontaneously* learn to echolocate or use it in their daily lives. In fact, there are actually some interesting and quite remarkable differences between blind people who are "expert echolocators" and those who do not actively echolocate at all. Not only is the ability to echolocate positively associated with certain measures of independence, the brains of echolocators also come to process echo information in ways that are strikingly different from how they process other sounds.[10]

In many ways, their brains come to treat the echo information as if it were visual information, which was also noted in the study I referenced in chapter four. Some of the more interesting findings in this area come from brain imaging studies, where the processing of echo information in the brains of blind echolocators takes place in the parts of the brain normally devoted to vision, something that is not seen in sighted controls. Even more interesting is the fact that when echolocation is used to identify the material properties of objects, such as whether they are made of plastic or fleece, for instance, expert echolocators show activation in a part of the brain involved in making the same type of distinctions through the visual channel. The brains of both sighted, and non-echolocating blind control subjects show no such pattern.

The same phenomenon appears when you study expert echolocators tracking a moving object, which then causes activation in a part of the brain normally devoted to visual motion processing. Processing of object shape in blind echolocators similarly causes activation of areas traditionally associated with the processing of shape by the visual system.[11]

One of the most striking findings concerning the difference between blind echolocation experts and those blind people who do not use echolocation has to do with how the former can make up for some of the typically expected deficits in spatial hearing.

As we learned in chapter four, the topic of spatial hearing is complicated, with blind subjects "overperforming" in horizontal space, and "underperforming" in vertical space, and when it comes to depth perception. Again, the reason for this appears to be that vision is needed in order to properly "calibrate" spatial hearing. However, one astonishing study has found that being able to echolocate can help compensate for a full loss of vision in this respect.[12]

Think about this for a second. One form of perception that is mediated by hearing can fill in for vision in calibrating *another* skill that is also tied to hearing. The sound coming from an active sound source and the echoes reflected back from an environment consisting of mostly silent objects are, at the end of the day, just sounds. They come in through the ears and go through the typical early sound processing that all sounds do. But, because of the different nature of these sources of sounds, and how they interact with the listener as he or she is moving or otherwise exploring the environment, the brain, at some point along the way, starts caring less about the nature of the *signal* and more about the nature of the *information* it can derive from that signal.

It suddenly makes perfect sense for the visual brain to step in since echo information is able to provide the brain with the particular type of knowledge about the world that we would ordinarily associate with seeing. No wonder so many blind echolocators through the centuries have insisted that what they are doing is *not* hearing, but something else entirely. And no wonder *Daredevil's* creators were so confused about what to make of the radar sense (much more on that later!).

ECHOLOCATION ON STEROIDS

Most modern accounts of echolocation focus on the likes of Daniel Kish and the late Ben Underwood who use tongue clicks to generate an optimal signal for detecting objects.[13] Much of the research literature on the topic from the last twenty years has also looked specifically at this particular method of generating echoes, including which sounds echolocators tend to gravitate towards (they typically feature a wide range of "high(ish)" frequencies between 3,000 and 8,000 Hz). Bat echolocation, too, relies on the animal actively producing a shriek. And even though Matt actually snapped his fingers on a couple of occasions during the *Daredevil* show, it was under special circumstances and not part of some broader pattern.

But if Edward Smallwood and other historical echolocators are any indication, people *can* obviously navigate without such deliberate techniques. There are also other intriguing sources of ambient sound that provide reliable sources of guidance. One such example is actually mentioned in the Dallenbach paper when describing Supa and Smallwood: "As they both reported, their perception of the side walls enabled them to follow the path to the end wall that was equally distant from the sides."

The phenomenon we're given a glimpse of here was investigated in the late 1990s in two papers that noted that the "non-uniformity" of sound pressure in a room may be one of the cues that facilitates navigation for people who are blind. Specifically, the sound pressure is slightly higher near walls and other large surfaces to a degree that can be perceived by a person approaching a wall or moving alongside it. A study of a group of children from a school for the blind supported the hypothesis.[14]

The strength of this phenomenon varies greatly by frequency however, so if you stand some distance from a wall, the background noise on the side of your head that is closest to the wall will have a slightly higher pitch than the background noise on the other side. If you instead walk toward a wall, you might notice a rise in pitch as you approach. The problem with high-frequency sound is that the buildup starts (and ends) too close to the wall to be useful. Only at frequencies under 500 Hz is there a change in sound far enough away from the wall, 1-2 meters (3-6 feet), for it to be useful for navigation.

While lower sound frequencies have very long wavelengths and cannot reveal small features, I would still argue that the above finding has the potential to make heard differences in the ambient sound field much more useful to Matt Murdock than it would be for any real human. Heightened low-frequency hearing would allow him to easily detect the presence of larger objects and architectural features. Such a skill would be even further enhanced by a heightened ability to notice small differences in both intensity and frequency.

If we add to this an expansion of hearing into a higher than normal frequency range, and a slight reduction in the hearing threshold, the combined effects would naturally provide much more information by which to detect objects and surfaces by ear than what is available to people in the real world. High-frequency sounds make for great echoes since they reflect well off of surfaces and have shorter wavelengths that can reveal smaller features of the physical environment. High-frequency sounds are also easy to locate using the interaural level difference as the head itself effectively shades sound coming at it from an angle, creating detectable differences in volume between the two ears. (Again, this of course also illustrates one of the pitfalls since higher frequencies don't travel very far.)

In the book *Spaces speak, are you listening: experiencing aural architecture,* by Barry Blesser and Linda-Ruth Salter, we find an excellent summary of the many subsets of skills that contribute to the totality of sound-based navigation:

> "Human echolocation is actually a collection of independent abilities to perform a variety of tasks, from hearing spectral changes produced by a nearby wall, to hearing the acoustic shadow produced by a telephone pole, to hearing the reverberation arising from two coupled spaces. A given listener might be very good at one task but mediocre at another."

Assuming that Matt's hearing would be better on every single measure – hearing range and threshold, as well as finer temporal, frequency, and intensity resolution – he would be well equipped to excel at each of these subtasks, especially if we provide his brain with the best conditions imaginable for making sense of the stimulus.

Whether we suppose that Daredevil's "radar sense" is its own separate sense or a heightened form of echolocation, explaining how he would detect the presence and shape of objects is much less of a challenge than explaining how he can hear faint sounds from great distances and is ultimately one of those things that makes him the perfect character for illustrating what is *almost* possible.

CHAPTER 7
THE FORGOTTEN NOSE

" "I know by smell the kind of house we enter. I have recognized an old-fashioned country house because it has several layers of odours, left by a succession of families, of plants, perfumes, and draperies. [...]

Out of doors I am aware by smell and touch of the ground we tread and the places we pass. Sometimes, when there is no wind, the odours are so grouped that I know the character of the country, and can place a hayfield, a country store, a garden, a barn, a grove of pines, a farmhouse with the windows open."

The World I Live In, by Helen Keller (1908)

In *The Man Who Mistook His Wife for a Hat and Other Clinical Tales,* the late British neurologist Oliver Sacks tells the strange tale of 22-year-old medical student "Stephen D." In the chapter *The Dog Beneath the Skin,* we learn of his experiences following experimentation with a cocktail of recreational drugs. The story begins:

"Vivid dream one night, dreamt he was a dog, in a world unimaginably rich and significant in smells. ('The happy smell of water... the brave smell of a stone.') Waking, he found himself in just such a world. 'As if I had been totally colour-blind before, and suddenly found myself in a world full of colour.' [...] But it was the exaltation of smell which really transformed his world: 'I had dreamt I was a dog – it was an olfactory dream – and now I awoke to an infinitely redolent world – a world in which all other sensations, enhanced as they were, paled before smell.'"

Stephen D goes on to describe how he could now go into a scent shop, and distinguish each scent immediately. Similarly, he could distinguish his friends, and his patients, by smell alone:

"I went into the clinic, I sniffed like a dog, and in that sniff recognised, before seeing them, the twenty patients who were there. Each had his own olfactory physiognomy, a smell-face, far more vivid and evocative, more redolent, than any sight face."

This story made a profound impression on me the first time I read it, while on an Oliver Sacks binge in my early twenties. It was made all the more intriguing by Sacks' admission, many years later in an interview about his 2012 book *Hallucinations*, that the young "Stephen D" had in fact been the great neurologist himself.[1]

Long before learning of Sacks' own experiences with hallucinogenic drugs, what had originally made the story of Stephen D so compelling were the unanswered questions it presented us with regarding the role of olfaction in human life. Of course, the experience as described must have been largely hallucinatory in nature. When Sacks (Stephen D) describes experiencing the spatially distinct scents of many different people when entering a room, it rouses the suspicion that his brain is activating the sense of smell in response to something in the visual channel and somehow merging the two senses.

Still, I couldn't help wondering whether there was something more going on here. Had something about the mix of cocaine, PCP, and amphetamines altered the young medical student's brain in such a way that a veil was lifted? Is there something *other* than just the quality of our noses that cuts us off from the mysterious scent scape that we know to be available to other animals? Why is it that scents don't usually – with the exception of strong or offensive odors – seem to force themselves on us, and grab our attention the way sights and sounds do?

Of course, if we look back at the historical record, this was hardly the question being asked by scholars of centuries past. They instead seemed to have settled on the human sense of smell being worthy of its low status. The 19th-century anatomist Paul Broca – perhaps best known for his discovery that a kind of aphasia could be linked to damage to a particular part of the brain – divided animals into groups of microsmatic ("poor smellers") and macrosmatic ("good smellers"). Humans and other primates were relegated to the former group. With this in mind, it is perhaps not surprising that we have tended to associate the sense of smell with "lower" animals. The father of psychoanalysis, Sigmund Freud, definitely had his mind made up. In *Civilization and Its Discontents*, Freud writes (emphasis mine):

 "The diminution in importance of olfactory stimuli seems itself, however, to be a consequence of man's erecting himself from the earth, of his adoption of an upright gait, which made his genitals, that before had been covered, visible and in need of protection and so evoked feelings of shame. Man's erect posture, therefore, would represent the beginning of the momentous process of cultural evolution. The chain of development would run from this onward, *through the diminution in the importance of olfactory stimuli* and the isolation of women at their periods, to a time when visual stimuli became paramount, the genitals became visible, further till sexual

excitation became constant and the family was founded, and so to the threshold of human culture. [...] It would be incomprehensible, too, that man should use as an abusive epithet the name of his most faithful friend in the animal world, if dogs did not incur the contempt of men through two of their characteristics, i. e., that *they are creatures of smell* and have no horror of excrement, and, secondly, that they are not ashamed of their sexual functions."

For anyone who might be interested in how all of this, in Freud's mind, ties in with children's relationship with their own feces and "anal eroticism"(!), I recommend a complete reading of the original text. However, what matters for our purposes here is not so much man's relationship with his genitalia, as Freud's view of the sense of smell in the context of evolution and civilization. On this topic, his position is clear. Civilized adults clearly have little interest in the world of scent, and for good reason!

Of course, Freud's lack of appreciation for the sense of smell may have been of a more personal nature. In the book *What the Nose Knows*, psychologist and scent specialist Avery Gilbert suggests that Freud himself likely had a reduced sense of smell owing to his medical history which included a severe case of influenza, persistent nasal congestion, the cocaine he snorted to treat his migraine, the cigar-smoking habit, and two rounds of surgery to his nasal cavity. It's not difficult to imagine that he would be under the impression that the world of smell is one we are more intimately acquainted with as children, before moving on to bigger and better things, when this neatly parallels Freud's personal experience.[2]

While Freud doesn't carry all of the blame for our disrespect of this sense, the view of smell as somehow beneath us can be encountered elsewhere. If we return to David Bolt's paper on *Beneficial Blindness*, we find an animalistic portrayal of this sense among the common literary tropes associated with the blind. Bolt notes the depictions of this sense in works such as José Saramago's *Blindness*, where the blind patients are seen "twitching, tense, their necks craned as if they were sniffing at something, yet curiously, their expressions were all the same." Another group of blind characters "stopped, sniffed in the doorways of the shops in the hope of catching the smell of food."

YOUR NOSE IS BETTER THAN YOU THINK

In more recent years, the tide has turned. Now we are being informed by research, communicated in the popular press, that humans are not nearly as bad at detecting smells as we thought we were.

You may recall having seen images in the news of blindfolded study subjects on all four, successfully tracking a scent along a path on a lawn. If so, you can most likely trace it back to a 2006 study conducted at the University of California, Berkeley, where thirty-two undergraduate students were recruited to put their scent-tracking abilities to the test by following a piece of twine dipped in a chocolate solution.[3] Most of the students were able to track the twine all the way to the end. While no match for a dog in this particular exercise, it turns out that we humans are not actually terrible at it. The students also improved with practice.

Other findings reveal that we can discriminate among an astonishing number of different odors. You may have heard the number 10,000 mentioned, but this is one of those numbers that are given the ring of truth by virtue of being repeated enough times by seemingly authoritative sources. In *What the Nose Knows*, the previously mentioned Avery Gilbert goes on an ambitious mission to find out where this number originally came from and manages to trace it to a late 1927 "guesstimate" made by the little-known chemical engineer Ernest C. Crocker. The real number of separate odors we can distinguish will probably continue to elude us, but is sure to be much, much greater. A 2014 paper in the journal *Science* gave an estimate of at least one *trillion*.[4]

Our noses are also sensitive to sometimes vanishingly low concentrations of many scents. The detection threshold varies depending on the substance but can be as low as a few parts per trillion(!), or the equivalent of mere drops dissolved in the volume of an entire swimming pool. Having worked in a biochemistry lab myself, I distinctly remember the pungent chemical β-mercaptoethanol which, among other things, is used to extract RNA from cells. It is also always handled in a ventilated hood, due to its toxicity. In spite of these precautions, the pungent smell (I would describe it as that of a damp, "mushroomy" bog) is unmistakable, and one I am sure I would recognize immediately.

The odor threshold for β-mercaptoethanol is about one part per two million, but other so-called thiols – a class of sulfur-containing organic compounds – can be detected at much lower levels. The thiols pop up in all kinds of contexts, from the hard-to-pronounce 3-methyl-3-sulfanylhexan-1-ol, a component of human sweat, to the furan-2-ylmethanethiol that is a key component in the smell of roasted coffee. Another thiol, ethyl mercaptan, is added to otherwise odorless natural gas to warn people of gas leaks. It has a detection threshold of less than *ten parts per trillion!*

But if our noses can do all this, why is it that we need to be told that our sense of smell isn't "all that bad"? And why did the takes of Broca and many others seem to track so well with our own experiences? Imagine if the newspapers ran stories declaring that "You hear better than you think!" Most people would probably be baffled by such a suggestion. The average person with normal hearing is unlikely to be thinking that her ears are under-performing in some crucial way, and the ubiquitous world of sound doesn't soon let us forget it. We know that there are critters around us who can hear some of the sounds that we cannot but that, as we have learned, also works the other way around. What we *can* hear appears to more than meet our needs.

For some reason, our sense of smell seems different, more elusive.

When we're not cooking, emptying the trash, or leaning away from the odd-smelling fellow next to us on the bus, we tend to forget about our noses. We look at our house pets with amusement as they explore a plain-looking patch of sidewalk, and accept the idea that the family dog experiences the world of scent in ways that are mysteriously inaccessible to us.

In a 2010 review article in the journal *Experimental Brain Research*, its authors Lee Sela and Noam Sobel proposed a theory for what might cause this discrepancy between the very low detection thresholds that are reported for human olfaction, and the way we detect scents (or not!) in our daily lives. They note that while the previously mentioned ethyl mercaptan can be detected at vanishingly low concentrations in experimental settings, it is added as a warning agent to propane gas at a concentration that is *57,000 times higher* than its detection threshold, to make sure we don't miss it.

Sela and Sobel point out that we humans are good at directing our attention to both sights and sounds in three-dimensional space. But, as we've learned, we are not as able to readily associate scents with a discrete location, which makes them less of a "thing" that can be spatially attended to. Other animals are better spatial smellers than we are, even though the stimulus itself is also a little trickier to contend with (as noted in chapter two).

When it comes to temporal changes, that is changes that occur over time, we are used to receiving continuous input from both our eyes and ears and are primed to notice when something changes. But even here, we are susceptible to a phenomenon called *change blindness* which makes it difficult to detect a small change between two otherwise identical stimuli when there is a brief pause between them. For instance, imagine that you are presented with two nearly identical and moderately complex photographs of a scene, one after the other, and have to spot the difference between the two. Unless the difference is particularly glaring, this may be quite difficult to do. On the other hand, if you were presented with these two images in a way, digitally say, that allowed you to see the progression from the first to the second image, the change would be more readily apparent.[5]

In vision and audition, the phenomenon of change blindness (or "change deafness") is more the exception than the rule, but when it comes to olfaction, Sela and Sobel argue, the situation is reversed, predisposing olfaction to its own kind of change blindness, or "change anosmia" rather. Even when we actively engage with odors, we do so through sniffing behavior which by its very nature is not a continuous activity but one that delivers the stimulus in discrete bursts.

The very fact that sniffing plays a major role in olfaction also underscores the importance of our active participation. While strong odors *do* grab our attention – and there are certain smells most of us would prefer to ignore if we could – the detection of more subtle odors may well require that we actively try to sniff them out. I have tried this out myself in situations where I don't think I am smelling anything in particular and have been surprised by what my nose picks up when I tell it to do its job. I have also come to be quite confident in my ability to smell when food has gone bad by making a habit of actively familiarizing myself with what various edibles smell like at their freshest. Smelling ability is highly amenable to practice. Actively training your nose, something that professional perfumers or sommeliers have to do, leads to a reorganization of the olfactory areas of the brain.[6]

Whether "change blindness" and other factors having to do with attention and behavior fully explain our lack of confidence in our sense of smell they seem to me to provide a decent explanation for why it is so often forgotten. Clearly, the difference between a human and a dog, in relation to smell, goes beyond the nose itself into the realms of behavior. Dogs are naturally compelled to sniff and to orient their lives around the scents they detect in ways that we are not. I don't mean to downplay the significant anatomical differences between dogs and humans, but am merely pointing out that these seem to go hand in hand (or should that be "nose in brain"?). In her book *Inside of a Dog*, author Alexandra Horowitz contrasts the olfactory worlds of dogs and us, noting that:

> "Not only are we not always smelling, but when we do notice a smell it is usually because it is a good smell, or a bad one: it's rarely just a source of information. We find most odors either alluring or repulsive; few have the neutral character that visual perceptions do. We savor or avoid them. Our own weak olfactory sense has, no doubt, limited our curiosity about what the world smells like. [...] As we see the world, the dog smells it. The dog's universe is a stratum of complex odors. The world of scents is at least as rich as the world of sight."

However, just because we don't consciously detect everything our noses pick up doesn't mean that we don't do so *subconsciously*. By now, everyone knows that one way to make your home more attractive to interested buyers is to bake something just before the viewing – in my part of the world, cinnamon rolls are a natural choice – or make fresh coffee.

Scents have been demonstrated to influence behavior in other domains as well. The smell of detergent can inspire college students to clean up after themselves even when they don't consciously notice the scent, and people are even susceptible to the smell of fear.[7] We are even compelled to smell our hand after we've shaken hands with someone, or touched something.[8] Not only are we largely unaware of what we smell, but we also appear to be almost equally unaware of some of our own *behavior* around smell!

I think it's interesting to consider that Daredevil may not even need a huge bump in olfactory abilities as such. Just being more consciously tuned into this sense, which might come about by changes to his brain rather than to the nose itself, would make a huge difference. Lift the "veil," and quite extraordinary things may follow.

· · ·

But there is also another way to smell at which humans excel, and that is the detection and appreciation of *flavor*. I mentioned earlier that when we talk about our sense of taste, we usually fail to distinguish it from the experience of flavor which is actually primarily about smell. When we detect "flavor-smells," we do so through the back of the nose, or through the *retronasal* pathway.

If you suspect that I am mentioning this as some kind of consolation prize, think again. The anatomy of our nose and airways alone suggests that we are actually *specialized* for retronasal smelling. At some time in the distant past, humans and other primates lost a piece of bony anatomy that effectively separated the mouth from the nose and restricted the exchange of air between the two cavities. Most other animals don't have these adaptations for retronasal smelling; dogs, for instance, will sniff their food and gulp it down, but they don't savor their food the way we do.

In *Smellosophy*, A. S. Barwich compares the experience of a dog to that of drinking coffee. Coffee is characterized by a strong and very alluring smell, whereas *drinking* the coffee is less exciting. In the case of coffee, the "orthonasal" experience, i.e. the smelling through the front of the nose, is the more satisfying one. This makes coffee an excellent example of a substance that elicits very different experiences depending on the route it takes. Cheese is another one, where the stinkiest of cheeses can create a much more innocuous and pleasant *eating* experience.

Our specialization for retronasal smelling explains why every human culture on the planet makes a big deal out of eating. Both cooking and the use of spices and herbs are close to being human universals, and the meal is a pleasurable experience to be shared with friends and family, not merely a source of nutrition. Where dogs obsess over who else was at the fire hydrant, we spend copious amounts of time and money on gastronomic pursuits, all while neglecting the central role that our sense of smell plays in all of it.

THE SCIENCE OF HUMAN SMELL

We have already noted that humans have a pretty decent sense of smell, but one that generally figures less prominently in our conscious lives than it does in the lives of certain other animals, at least when it comes to orthonasal smelling. But what do the nuts and bolts that convey smells actually look like?

The science of olfaction still has a long way to go, and some of the key discoveries in the field have come relatively late. The discovery of olfactory receptor genes in 1991, by Linda Buck and Richard Axel, and their subsequent work on the olfactory system, was considered groundbreaking enough to win them the 2004 Nobel Prize in Physiology or Medicine.[9] Buck and Axel carried out their initial research on rats, but the new insights made it possible to compare the number and type of these genes between species.

When the spotlight turned to humans, the discoveries initially seemed to confirm that old notion of ours that humans were indeed on an evolutionary course to olfactory oblivion.[10] It turns out that the human genome contains around 800 olfactory receptor genes. If you've been following along and still remember some factoids from chapter three, you may remember that the proteins these genes code for are all variations on the G-protein coupled receptor.

You may also be thinking: Hey, weren't there supposed to be around 800 of these in *total*? With the ones devoted to smell accounting for about *half*? You would be correct. What gives? Well, it turns out that just over half of these genes are non-functional so-called pseudogenes. Pseudogenes are stretches of DNA that used to code for functional proteins at some stage in an animal's evolutionary history but have sustained mutations that prevent them from performing as they once did.

In essence, the average human has just under 400 olfactory receptor genes that actually work, in the sense that they code for functional receptor proteins. Dogs, on the other hand, have more than 800, mice more than 1,000, and rats over 1,200. On a more positive note, we at least fare significantly better than chickens and platy-puses in this comparison.

What is the practical significance of the number of olfactory receptor genes? Well, first of all, we should note that *every* species examined has a fair number of non-functional pseudogenes, in addition to their functional genes. The fraction of pseudogenes as a proportion of the total number of identifiable olfactory receptor genes is significantly higher in primates (~50%), but even dogs, mice and rats carry around a fair share of genetic junk (~25%). This is a natural consequence of the dynamic history of this particular family of genes, where new gene variants have continued to evolve while others have been lost. The end result of this process is that the olfactory receptor gene family has come to be very diverse, with relatively little overlap between species.

Earlier research seemed to indicate that the apparent reduction in the size of the repertoire of olfactory receptor genes in humans, other apes, and most of the Old World monkeys might be a consequence of the evolution of full-color vision. We humans and our close ancestors are so-called trichromats, whereas most other mammals are dichromats. The idea that the evolution of trichromatic color vision caused the concurrent reduction of functional olfactory receptor genes in these primates is known as the vision priority hypothesis.[11] With improved color vision, the selection pressure that served to keep many of the olfactory receptor genes functioning lessened. At least that was the *idea*.

This too has turned out to be not quite right. A 2010 study found that there was no significant difference in the size of the repertoire of olfactory receptor genes between dichromatic New World monkeys and their trichromatic relatives, including us humans.[12] It was also established that humans retain more of the olfactory receptor genes that were present in the common primate ancestor than orangutans and macaques do. In this regard, humans are more similar to marmosets, which might come as a surprise to most people. Sigmund Freud would be shocked, I'm sure.

To make matters even more complicated, another study that

looked for the levels of expression of various genes in the olfactory epithelium found that about 60 percent of the olfactory receptor neuron *pseudo-genes* were actually expressed to at least some extent. The implications of this finding aren't exactly clear since this doesn't mean that they form a complete protein. But they might have some kind of function.[13]

Whatever the case may be, the relationship between the number of functional olfactory receptor genes and the olfactory prowess of an animal is not as clean-cut as the early findings might have indicated. We think of dogs as representing the pinnacle of smelling prowess, but their repertoire of functional olfactory receptor genes is not particularly large by typical mammalian standards, even though it's twice as large as that of humans. And the fact that we can distinguish between an almost impossible number of scents even with our relatively lower number of functional receptor genes makes you wonder if there would be much to gain by reawakening all of those pseudo-genes, even though it might certainly be an interesting avenue for Daredevil.

While we now know how our noses encode smells on a chemical level, the sense of smell is in many ways still a mystery. Each olfactory receptor neuron expresses only a single kind (though multiple copies) of olfactory receptor protein. When you look at the olfactory epithelium, where chemicals and noses meet, there is no pattern at all as to what olfactory receptor protein is expressed in which part of the epithelium. However, all neurons expressing the same protein appear to cluster together at the next stage of processing, the olfactory *bulb*. This is where we can look for that combinatorial code I mentioned in chapter four. From the olfactory bulb, the information is sent to several different areas of the brain, though notably not through the thalamus like the other senses. The primary olfactory cortex is located on the inside of the temporal lobe (that is, on the inside of the side of the brain that looks like an earmuff). Here, the information is consciously perceived and becomes available for integration with the other senses.

But knowing this still doesn't tell us much about what kinds of things smells are, as perceptual objects. There also doesn't seem to be a particularly close relationship between the chemical structure of a particular scent molecule and whether it binds to a particular sensory receptor. A stable relationship between molecules and receptors exists, but knowing about the chemistry of these molecules isn't really helpful and doesn't predict their perceived scent quality. I suggest we all make sure to keep an eye out for new developments in scent research in the coming years. While the current mystery surrounding olfaction can be frustrating, that also means that much remains to be discovered.

THE SMELL OF HAIR TONIC

The cluelessness we seem to have about our own abilities in this area may explain some of what has been going on in the comics. Apparently, our sense of smell is so forgettable that even many of the writers of *Daredevil*, especially the early ones, could go several issues without so much as a mention of the main character smelling anything interesting at all; this in spite of the fact that a heightened sense of smell usually ranks high on lists of Daredevil's superpowers. Heck, Matt doesn't even appear to be *tasting* much of anything so we can't even really tell whether he gets an unusual kick out of the flavor of food, or if his heightened sense of smell is restricted to what's coming in through his nostrils. In *Daredevil* #1, Matt Murdock gives the reader this modest description: "And I never forget an odor once I smell it! I could recognize any girl by her perfume... or any man by his hair tonic..."

Not only does this not sound particularly impressive – we are, after all, presumably dealing with the realm of superpowers – the only thing Matt's nose is actually used for in the origin issue is to track the Fixer's cigar, a scent so strong and distinctive that any perfectly average person would be unlikely to miss it. Over the course of the following twenty issues, Daredevil's sense of smell is refer-enced, on average, less than once per issue, with most instances being fairly mundane. There is one mention of Karen's perfume, both the Owl and the Ox apparently use hair tonic, and there are a couple of references to the smell of gasoline.

What strikes me as the most creative example of the bunch is a line from *Daredevil* #18, by Stan Lee and John Romita: "I smell salt water, and the kind of raw lumber used in packing crates! We must be near the docks!" Here, we actually have Matt noticing useful pieces of information about his surroundings, and doing so in a way that doesn't shy away from the core concept that his remaining height-ened senses are supposed to be relaying vital information about things he cannot see. As mentioned though, this example stands out to us because it is rare.

It also seems that Daredevil repeatedly fails to live up to his own claim that he can identify people by smell. Or, if he can, he doesn't seem to be doing much of it. When Matt encounters Peter Parker in court, in *Peter Parker, the Spectacular Spider-Man* #107 (1978), he recog-nizes him by his voice and his heartbeat, which is highly typical of what we have come to expect over the years. By the time they reveal their identities to each other at the end of the Death of Jean DeWolff storyline, in issue #110, the only mention of smell is in reference to Peter's "acrid" apartment, apparently the result of a recent fire in the building.

With the olfactory prowess you might expect from a *blind hero with a super nose* being nearly absent, the other senses are pushed to the forefront in distinctive ways that have shaped our view of the character. If Daredevil's creators hadn't been so quick to forget about Matt's sense of smell, would the *Daredevil* archives be quite so full of references to Matt distinguishing between people based on the sounds of their beating heart? I doubt it. It's not that the idea of using heartbeats as a biometric marker is *completely* nutty – see chapter five – but it would make much more sense for him to identify people primarily by smell than to do so by listening to their heartbeats.

What makes for a serviceable lie detector (*maybe*) doesn't necessarily provide the best way of distinguishing between people.

Failing to take smell into account can also introduce unfortunate plot holes. In the first season of the *Daredevil* show, as Matt tends to Claire's wounds in his apartment in episode five, "*World on Fire*," he notices that the cut has opened up by the "taste of copper in the air." However, in the first episode of season two, "*Bang*," Matt fails to notice that the character Grotto is bleeding profusely until the latter almost falls off his bar stool. In season three, we are to go along with the fact that Matt can detect the faint traces of his Daredevil costume in Dex's apartment, but that he at no point connects the smell of the "new" Daredevil with the FBI agent he met, and spoke to at some length, three episodes before their more violent encounter.

Expecting that Matt would actually be able to reliably identify everyone he's ever met by smell makes for challenging storytelling and calls for a degree of infallibility of human memory and cognition that strains credulity. But given everything *else* Daredevil can do, we probably should have expected Dex to at least appear vaguely familiar to him when they battle it out in the sixth episode, "*The Devil You Know*," especially since he wouldn't be thrown off by the *sight* of the guy.

That said, I really appreciated the way the scent of a particular dry-cleaning establishment became a plot point during the beginning of *Daredevil's* third season, as the scent allowed Matt to make a connection between some bad guys and their legitimate business.

Matt's nose would eventually get a little more respect from the comic book creators as well, and if you sit down and read every issue of Daredevil in chronological order – as one obviously should on a semi-regular basis – it does eventually become more prominent, especially during the 70s, and Frank Miller used it quite heavily. *Daredevil* #168 (1964), the first to list Frank Miller as both writer and artist – with Klaus Janson on inks – opens with:

"Close your eyes, let the night touch you. *Feel* the cold, driving rain as it batters your face and soaks your clothes... *Hear* the moan of a freight barge on the nearby East River: The haunting chimes of a solitary church bell as it tolls the midnight hour... *Taste* air heavy with lingering fumes of rush-hour traffic long gone... *Smell*, in maggot-ridden garbage, the stench of another day's misery in New York's Lower East Side... Let the night touch you — and you will take in only a *fraction* of its total texture... a texture fully experienced by only one man — a *blind* man."

The description Miller offers us has a more intimate and visceral feel to it that was rare during the Silver Age era and, importantly for the point I'm trying to make here, includes a generous treatment of the sense of smell (heck, there's even some taste thrown in). The smells wafting their way up to Matt Murdock's nose may not be particularly pleasant, but that only serves to ground the character in previously unrivaled levels of grittiness.

This doesn't mean there are not some honorable mentions of useful smelling among the earlier writers that are worth mentioning. In *Daredevil* #46, by Stan Lee and Gene Colan, Matt identifies a white coat by its freshly-laundered smell and puts it on to impersonate a doctor. In *Daredevil* #146, by Jim Shooter and Gil Kane, Matt identifies a gun shop by the "[s]mells of cordite and bluing!" Cordite would go on to become something of a Daredevil staple, even though production of cordite ceased sometime after World War II, and is not something he ever should have been smelling.

Some of the very best scent writing to date came during Ann Nocenti's run, and she even dared to have Daredevil noticeably sniff the air, in ways we presume were not much too "animalistic." In an early issue, *Daredevil* #238, where our hero goes up against the villainous mutant Sabretooth, the latter actually clocks Daredevil as blind after he catches him exhibiting some unusual behaviors. "What's he doing...? He sniffs... Tilts his head... *He's not using his eyes!*"

In *Daredevil* #243, Matt pieces together the identity of his unusual voodoo-practicing opponent: "Other smells -- tree sap... clay... oil... musky paint on his skin... He's not American, perhaps from some tribe? Haitian like Danny? He must be one of Danny's men!" In *Daredevil* #250, it is someone else's identity that becomes the object of Matt's nose: "What's this? The polished smell of imported leather... The finest tobacco aroma... The arrogant stance, aristocratic tilt of the head... the over-starched shirt...What's an uptown job like him doing *slumming* in a *free* clinic in *Hell's Kitchen*, the nastiest neighborhood in New York?"

Interestingly, Nocenti also makes a couple of mentions of Matt detecting the smell of tears. In *Daredevil* #239 he notices something while passing a scene on the sidewalk: "What's the commotion? Smell of new clothes... Rice -- a wedding! Hmm... The salty smell of tears... Tears of joy." In Daredevil #240, we catch him studying Karen:

> "The red man is blind. His sight was robbed by a radioactive isotope during a childhood *accident*... that was really a *gift*; for it super-heightened his remaining senses and even gave him a compensating radar. He can't *see* the woman he loves, but he can listen to her big heart beat, he can smell her sad salty tears when she cries, feel her anger when it ripples across her skin, sense the warmth of joy when it rises in her body... and somehow, he even knows the moment her lips curve into a smile."

It seems doubtful that it is the "saltiness" he's smelling, if he is smelling anything at all, and I would be doubtful of the tears from the wedding providing such a noticeable stimulus against that kind of background. However, in a 2011 study published in the prestigious journal *Science*, a group of researchers – including the previously mentioned Noam Sobel – studied men's reactions to women's tears and turned up some astonishing results![14] The men in the study couldn't actually *smell* the tears, or consciously tell them apart from a drop of saline that had also touched the women's skin. However, exposure to the tears – but not the saline solution – lead to a reduction in sexual arousal. This is the kind of thing I would naturally be inclined to be skeptical of, but the design of the study was fairly robust and was also double-blinded. That means that not even the experimenters knew whether the subjects were exposed to the real tears or the controls. I guess we will have to wait and see whether anyone else is able to replicate these findings!

THE SCENT OF A MEMORY

What about the connection between memory and smell? What is the nature of the connection, and is it true that memories tied to scent are more closely tied to emotions than other memories?

At this point, I am of course obliged to bring up Marcel Proust's tea-soaked madeleine which famously transports him back in time to childhood memories of his aunt in *Remembrance of Things Past*. One could be forgiven for believing that Proust introduced the connection between memory and scent to the world of literature, with all the attention this scene still gets, but Avery Gilbert again reminds us that this is not the case. Nor does Proust's recollection seem typical of the experience we usually have when scent triggers the memory of something long ago or far away, in that the memory does not come to him quickly but through strenuous deliberation. It is also telling that we don't learn much about the scent experience itself, or whether Proust even enjoys madeleines.

As Daredevil fans, we can happily look to another story for an example of the occasional richness of smell, and how much this sense means to Matt Murdock. The story arc *The Devil Takes a Ride*, by Ed Brubaker and Michael Lark, covers *Daredevil* #89 (1998) through #93 and sees Matt manipulated by a Lily Luca who we later learn is a pawn for the villainous Mister Fear. Lily is wearing a very special perfume that affects Matt and brings out his protective instinct while causing him to lower his guard.

In *Daredevil* #91, as Matt realizes he has literally been (mis)led by the nose, we are treated to a special look at the sense of smell. Matt recalls the time following his accident and the scents that assaulted him on the streets of Hell's Kitchen. He also remembers the strong link between smell and his father's memory:

 "[...] And people each came with their own distinct scent, whether they realized it or not. Like my dad. No matter how much he washed, he could never get rid of the smell of the gym... The heavy bags, the leather gloves, and that strange *heavy* air... a combination of smoke and sweat and humidity... No one but me would notice, but that was the scent of my father, Battlin' Jack Murdock. He smelled like a fighter... even when he was on a slab in the morgue. And in our old apartment, after that, the smell was inescapable. It seemed like he was everywhere, from the cupboards to the pillow-cases... That first night, I slept in his closet. And when I awoke, the sense of him was so strong, it was like he was still there... just for a few moments. That was when I realized how cruel my senses could be... Because scent triggers things in the brain that you can't control... Feelings and memories... of better times."

As smell remains an oft-forgotten sense, at least in relation to its potential significance and usefulness, this scene serves as a beautiful reminder of what smell can potentially do for all of us, and for Matt Murdock in particular.

I remember a relatively recent visit to my mom's house when she took out a cardigan she had kept from my grandmother's belongings following her death in 2005. We both remarked that it still smelled like my grandmother! There is something almost intimate about smell, in that the scent of a lost loved one is quite literally physical traces of the person, while a picture or a sound-recording of their voice is not.

·   ·   ·

There exists a kind of folk mythology around smell and memory, and there are even scientific findings that at first appear to back up these claims. As we noted earlier, the input from our noses doesn't pass the thalamus the way our other senses do and go straight to the higher areas of the brain. This has been alleged to imbue scent memories with layers of emotional content that other sensory memories lack.

While you will occasionally see a report pointing in this direction, perhaps particularly in the popular press, the totality of the science doesn't support that our memories of smell are better or more accurate than memories associated with the other senses, or more immune to the passing of time. For every study purporting to reveal something special about smell, there are others that find nothing of the sort.

When it comes to whether there is a greater emotional component to smell, the answer seems to depend on who you look to for guidance. In *What The Nose Knows*, Avery Gilbert points to the surprise factor as something that contributes to our reaction to smell-induced memories. Owing again to the fact that we appear to be smelling so many things that we are not *consciously* aware of, it naturally comes as a great surprise to us when we recognize a scent we didn't know we had stored away to begin with, and are nonetheless able to associate with something meaningful.

Other experts do seem to think that certain such memories, while not necessarily more emotional, have a special quality to them. To quote A. S. Barwich:

"What makes odor memory mesmerizing is the feeling that it sometimes transports you, almost physically, to a different place or time. Smells evoke presence, and immediate physical embodiment."

We could potentially account for almost everything that Daredevil has been smelling throughout the history of the comic by simply connecting him more closely to his sense of smell, behaviorally and cognitively, without doing much to his nose. Especially since it has been so criminally underused that a heightened sense of smell is rarely needed to explain anything. Having said that, we could of course do to Matt's sense of smell what I suggested we do to his sense of hearing: Give him a greater number and density of olfactory receptor neurons in the nose, and more processing power in the brain. However, putting all of this to work for Daredevil might require that his creators start sticking their noses in spice jars and garbage cans before sitting down to write about this otherwise oh-so-easily forgotten sense.

CHAPTER 8
HEAT AND TOUCH

 "When I close my eyes and glide my hand over my writing paper, blotter, writing pad, and the cloth cover on my desk, I experience four distinctly different tactual impressions, one after the other. I feel the hard smoothness of the writing paper, the fibrous suppleness of the blotter, the leathery brittleness of the writing pad, and the soft roughness of the cloth cover. Once having become alert to these sorts of differences, if I now touch everything within reach, then I find that I not only can recognize the metal of the paper weight, the glass of the ink bottle, and the wood of the pen holder, but that I can even distinguish – naturally also with my eyes closed – between different types of paper by their surface features, which not even the eye can do."

The World of Touch (1925), by David Katz

There are some great stories in the history of science. From the world of medicine, we have Edward Jenner and his observation that milkmaids rarely came down with smallpox, due their exposure to the related but more innocuous cowpox virus. There is the equally compelling story of Sir Alexander Fleming, who in 1928 discovered penicillin when he came back from vacation and found an unwashed flask in which the bacteria it contained had been killed by the same mold that had been courteous enough to leave nearby traces of itself. Fleming's case is perhaps especially heartening for those of us who don't always keep our desks clean. Even a mess can herald a scientific revolution, saving millions of lives!

The story of how British astronomer Sir William Herschel (1738-1822) discovered infrared radiation – or as he would put it, "radiant heat" – is perhaps not as well-known to the general public, but just as compelling in its serendipity. In his studies of the sun, Herschel relied on glass filters of different colors and opacities to shield his eyes from the bright light. Attempting to find the best one, he noted that certain filters would heat up, but let little light pass through, and vice versa. This planted the seed in his mind that light beams of different colors might differ in their capacity to heat objects and surfaces.

In the early spring of the year 1800, Herschel set up an experiment in which he let light from the sun pass through a prism to separate out its constituent colors. Placing a thermometer underneath a slit that would only let a narrow band of light from the prism pass, he went on to measure the temperature across the range and found a rising trend from purple to red. Oddly enough, none of the colors of visible light seemed to produce the expected maximum temperature of an intact beam of light, which lead him to conclude that radiant heat consisted, perhaps entirely, of "invisible light."

What Herschel had discovered lay beyond what science could fully make sense of at the time, and the term "infrared light" wasn't coined until more than eighty years later. We now know that he had merely cracked open the door to the kind of wave phenomena, invisible to us, that make up most of the vast electromagnetic spectrum.

Herschel's discovery was obviously enlightening, and we will get back to it. However, we should first note that there still seems to be a great deal of confusion among the general public around the relationship between color, heat, and temperature as conceptual categories. We use temperature-related terms to describe colors in everyday speech – red is a "warm" color and blue is a "cool" one – but are we to take this literally? And does the color of an *object* affect how it feels to the touch, the same way particular colors of light from a prism differ on this point?

If you have read the early comics, you may suspect why I am bringing up this topic. Way back in the days of Silver Age comics, Daredevil used to be able to distinguish between colors by touch. In fact, this ability was used in the making of his classic yellow and black costume. To quote Matt himself, from *Daredevil* #1 (1964): "I'm no Betsy Ross, but I should be able to handle this! Lucky my touch is so sensitive! I can even blend the *colors*, for each colored fabric has a different *feel* to me!"

Of course, given that this particular color scheme was phased out after a mere six issues, and made Daredevil look about as frightening as a bumble bee, fans should probably have been questioning this ability from the outset.

Color-sensing was put to even more humorous use in *Daredevil* #60, in which Matt impersonates a thug by first ascertaining, also by touch, that the facial features of the unconscious man were sufficiently like his own, and then dyeing his hair to match as well: "Maybe I can only *feel* color… not *see* it… But a few hip-pocket *chemicals* will darken my hair-color to match *his*!" This scene of course raises all kinds of questions that have nothing to do with color, including: Does Daredevil *always* carry around mixes of hair dye to match any color?

Over the years, though, Daredevil's ability to sense color by touch has gradually been phased out. This makes it really the only early core ability to go this route.[1] By *Daredevil* #106, Matt had already started to doubt himself. In a scene where Daredevil has to help his new friend Moon Dragon push the right button, he is unable to do so. In his own words: "Moon Dragon... I–I'm *blind*! I can "see" colors with my *fingertips* — gauge the *heat* they absorb — but that takes time! And if I make one wrong guess!"

The last appearance of anything even remotely like color-sensing by touch can be found in *Daredevil #339*, which came out in 1995. What Matt is able to do here also comes closer to a more realistic scenario than any of the ones preceding it, so kudos to writer D.G. Chichester for giving this some thought. In the scene in question, Matt is wearing his new armored costume and is trying to pass himself off as a "new" Daredevil after having faked his own death about a dozen issues earlier. Confronted by journalist Ben Urich, one of the few people who knows that Matt Murdock and Daredevil are one and the same, he is given the same test that he failed years earlier. The caption reads:

"Make a *blind man* describe a *photograph*. It worked then. Murdock *can't afford* to let it work again. Light and dark areas *absorb* degrees of *heat* from the lamp above. Enough difference to paint a *crude picture* for hypersensitive fingertips. Some *deductive* guess-work on the photos a man might carry in his wallet. All adding up to enough for a *stab* in the dark."

There is one problem with this scene: light and dark areas do not differentially absorb *heat* from the lamp, they differentially absorb visible *light* specifically. This is a seemingly small but crucial difference. Visible light can absolutely heat objects, as we learned from Herschel's experiments, but the concepts are in no way synonymous. However, if we allow for that little caveat, there's at least theoretical ground to stand on here.

Since Matt has access to a lamp, and all he has to do is distinguish between lighter shades and darker shades, we might compare this scenario to what happens if you put a light and a dark object, that are otherwise identical, out in the sun for a while. A darker surface absorbs more of the visible light that hits it than a lighter surface, which is what makes it appear dark in the first place. The absorbed energy increases the heat of the object.

It would still be a stretch for Matt to be able to do even this (especially with his gloves on!), and the scene makes it clear that there's guesswork going on. But let us put the nail in the coffin for color-sensing by touch, once and for all.

COLOR, HEAT, AND TEMPERATURE

The very first thing to remind ourselves of where the topic of color is concerned is that we are talking about a visual phenomenon. The redness of red exists only in our brains, and it is an experience that can be had only through the visual modality. This, of course, doesn't mean that colors are arbitrary. When we look at a red apple, we know that it appears red to us by virtue of its surface characteristics. The skin of the apple preferentially absorbs the parts of the visible spectrum of light that our visual system designates as "not red," while reflecting much of the light that the same system understands as "red."[2]

With the knowledge that the surface characteristics of our red apple are stable, it may seem reasonable that it could provide us with at least an academic understanding of its color, by touch or some other means. Claims to that effect have actually been made, surprisingly recently and in scientific journals, by those who suggest that something called "dermo-optical perception" might be a real phenomenon. A 1992 paper in the journal *Perceptual and Motor Skills* states that "the basic assumption underlying the mechanism of dermo-optical perception is cutaneous sensitivity to infrared light emitted by colored surfaces." The study featured in this paper ultimately found no support for the phenomenon, however, and the authors concluded that "we can argue with some confidence that, if dermo-optical perception existed, we should have been able to measure it in some reliable fashion for some subjects at least."[3]

I was not the least bit surprised by this finding. I was more surprised by the suggestion that the colored surfaces might emit

different amounts of *infrared* radiation, consistent with their particular *color*. To explain why this doesn't hold up to closer examination,
and why detecting colors by touch is a Daredevil power that deserves
to remain buried, let us take a tour of some concepts from the field of
thermodynamics.

Let us first confirm that, yes, all objects we surround ourselves
with do emit infrared radiation. In fact, "thermal radiation" is emitted
by all matter in the universe with a temperature greater than absolute
zero, and the hotter the object, the more energy will be emitted. At
the temperatures we normally deal with in everyday life on planet
Earth, thermal radiation can be thought of as being more or less
synonymous with infrared radiation, which occupies the portion of
the electromagnetic spectrum just below visible light. However, with
increasing temperatures the peak wavelength of the radiation that is
emitted starts to skew lower (and the frequency higher), with greater
proportions of the total energy being emitted as visible light. This is
what happens in the case of a glowing hot iron, the flame from a fire,
or the sun. In fact, almost half of the thermal energy that reaches us
from the sun is in the visible part of the spectrum. Going back to
Herschel's experiments, what makes the sun an unusual object in our
lives is not that it emits infrared light, but that it is hot enough to *also*
emit *visible* light.

Since thermal radiation is emitted in proportion to an object's
temperature, we should also say a word or two about the related and
not entirely straightforward concepts of heat and temperature. *Heat* is
defined as the amount of energy flowing from one body (object) to
another spontaneously due to the difference in temperature between
them. Heat is more like a process than a stable property of an object.
When we touch an object we would describe as "hot," it is not the
case that the object "has" heat, but that heat is what flows from the
object to our hand. If we touch an object we describe as "cold," heat
flows *from* our hand to the object.

When there is no net transfer of heat between two objects, we
conclude that they are in a state of *thermal equilibrium*. When two
objects are at thermal equilibrium, they are said to have the same

temperature. In fact, this is how we measure an object's temperature with a thermometer—the measured temperature of an object is simply the value (in Fahrenheit, Celsius, or other arbitrary scales we've created) at which the thermometer is in thermal equilibrium with the object.

Thermal radiation is *one* of the ways that heat can be transferred between objects. Thermal radiation, like all electromagnetic energy, can move across great distances and through the vacuum of space, which is how we can be reached by the warm rays of the sun. This is not the only way to transfer heat, though.

Conduction takes place within an object, or between solid objects that are in physical contact. Through conduction, heat is transferred by the moving particles bumping into each other, in a kind of bumper car style. How quickly this happens depends on the material. Wood, for instance, is a poor conductor, and metals are very good conductors. This is why a metal object often feels (briefly) cold to the touch when at room temperature, as the bumper-to-bumper action in our (warmer) hand quickly "bumps" its way into the metal and cools the skin.

Finally, there is *convection*, which is the transfer of heat through gases and liquids. Gases and liquids are poor conductors because the particles they consist of are too far apart to be as good as solids at playing bumper cars with the next-door neighbor. Instead, a larger-scale process takes over by which there is a flow of particles through space. Most people are familiar with the idea that hot air rises to "float" on top of colder air. This is because gases expand with rising temperatures. This creates a mechanism for transferring heat that is a bit larger in scale. It's a bit like taking the bumper cars and leaving the country fair.

In reality though, conduction remains a contributing factor in these scenarios as well, since it's rarely the case that *no* conduction is taking place. The chilling effect of jumping into a cold lake – even though it may be the same temperature as the air! – comes from the fact that water is a much better conductor than air, and cools you down quickly.

. . .

Different materials do not only have different conductive properties. Surfaces also differ in the degree to which they absorb and emit thermal radiation. As you might recall from chapter two, visible light may either be transmitted, reflected or absorbed by the surfaces it falls on. The same logic applies to "invisible" infrared radiation. However, just like surfaces behave differently under visible light, which is why they appear to differ in color, they also exhibit specific interactions with other portions of the electromagnetic spectrum.

Surfaces that easily transmit visible light, such as a car windshield, may not be quite so willing to let infrared radiation pass through. This is why your car gets dangerously hot when you leave it out on a sunny summer day. The visible light from the outside is absorbed by the upholstery and other surfaces inside the car, which raises their temperature and increases the amount of infrared radiation that is being emitted. However, this radiation is mostly trapped inside the car. And yes, this phenomenon is also the basis for the greenhouse effect!

We can get a sense of how well different objects absorb and emit thermal radiation by comparing them to a "black body." In physics, a black body is an idealized imaginary object with a surface that perfectly absorbs and emits thermal radiation regardless of wavelength. The reason we think of it as black is simple, albeit somewhat misleading. In our visual world of colors, we register a surface as black if it absorbs all *visible* light. Surfaces that are white do the exact opposite and reflect the light. However, an object that appears black to us when we look at it may not appear "black" from the perspective of infrared radiation. For it to count as a black body it needs to appear "black" at *all* frequencies.

A black body defined this way is of course an abstraction, but we do determine the "emissivity" and "absorptivity" of real-world surfaces based on how closely they conform to the behavior of these extremes. A black body is defined as having an emissivity of 1 at all frequencies. Real surfaces fall on a scale between 0 to 1 which gives the fraction of how well they emit thermal radiation compared to a black body. Human skin is a very good emitter (and absorber) of infrared radiation, with an emissivity of roughly 0.98 at the relevant wavelengths. Notably, it makes no difference whether most of your ancestors hail from Ireland or Uganda – your skin color has no bearing on the emissive properties of your skin at these wavelengths.

The emissivity of a surface becomes a practical issue in thermal imaging and when trying to get a reading with an infrared thermometer which operates on the assumption that the emissivity of the particular object you're trying to measure is very close to 1. When this is the case, the measurement you get with an infrared thermometer will give you a very accurate reading of its true temperature. Polished metal surfaces, however, have a very low emissivity, and will appear to be much cooler than they actually are to this kind of thermometer. If you apply paint to such a polished metal surface, you will invariably raise its emissivity.

Does the *color* of the paint matter though? Because that is what the claim made by the proponents of dermo-optical perception ultimately boils down to. The short answer is no. The longer answer is that two differently colored paints may give different readings, but so may two black paints from different suppliers, depending on the specific formulation. If you throw other materials into the mix, you'll see the same thing. Two different brands and thicknesses of white paper will differ from each other. A red brick surface owes its thermal surface characteristics to its "brickiness" more than its redness.

Materials of different colors can certainly differ in ways that can be detected by touch, by whatever mechanism, but it is virtually impossible to control for the situation where the color difference doesn't stem from some *other* difference, such as the presence or absence of a particular pigment. There is no single quality connected to any color that makes it possible to identify it *in isolation* through touch alone. None.

In real-life situations, what an object feels like to the touch is going to depend on any number of factors, ranging from its texture, density, and conductive properties. It will also, obviously, depend on the temperature of the object. For everyday objects around the house, the temperature will be the same as the ambient temperature in the room, unless of course you just took something out of the fridge, or the oven.

If you do leave something out in the sun, or under a bright light, you will find that a dark object becomes warmer than a similar lighter object, but the difference will have more to do with how these objects reflect and absorb *visible* light than their much less predictable response to infrared radiation, whether from the sun or your hand.

Even though color-sensing has effectively been done away with in the Daredevil comic, there are other things having to do with the world of heat and thermal energy that occasionally come up in the comic and beyond. I have previously addressed the instances of *de facto* thermal imaging that were featured in the first season of the *Daredevil* television show, and why this simply could never work as portrayed.

This was not the first time in the character's history that an ability similar to this had reared its head. In *Daredevil* #74 (1964), by Gerry Conway and Gene Colan, there is one scene in which Daredevil is able to sense a flashlight in the alley below by its "infrared light," and similar scenes involving light sources specifically pop up from time to time.

Daredevil #377, by Scott Lobdell and Tom Morgan features a scene in which Matt (currently experiencing amnesia in France) claims that "I can feel the heat from their bodies, telling me who is standing where... while the sound of their heartbeats indicates who plans to strike first, and who is content to wait for the second wave assault." As we noted in chapter three, even *if* we were to equip Matt's skin with the ability to meaningfully detect infrared *as a form of light*, this would not form an image.

For a more classic and character-typical take on the sensing of heat, let us briefly go back to the 1970's and *Daredevil* #131, by Marv Wolfman and Bob Brown, an issue noteworthy for featuring would-be archenemy Bullseye's first appearance. Before Daredevil even gets to meet his new foe, he visits the crime scene of Bullseye's most recent victim. The police on site are not too happy to see him, but he makes himself useful by alleging that the murderer wore gloves, and that they need not bother looking for prints. How does he know? As we learn from his internal monologue, he has discovered an "absence of heat-residue on the pen" connected to a note found on the scene.

I think most readers instinctively sense why this proves nothing. Would Bullseye's handling of the pen raise its temperature? Matt is working from the assumption that the pen, originally at room temperature, could be expected to feel warmer to the touch as a result of any gloveless manipulation by a significantly warmer human hand. As assumptions go, this one isn't crazy. But as with other "stimulus problems," there are too many unknowns here. How long can we expect the pen to be unusually warm before once again reaching thermal equilibrium? Not very long. This time would have come and gone very quickly. And how do we know Bullseye didn't have his hand wrapped around a cold beer minutes before picking up the pen, completely thwarting the temperature reading? And what are we to make of the heat transferred to the pen by Daredevil's own hand? As a lawyer, Matt should know that this would never hold up in court. If you ask me, it would have made more sense to introduce Daredevil's consistently underused sense of smell here.

Another scene with some of the same problems appeared in the director's cut of the 2003 *Daredevil* movie where Matt can be seen reaching up to touch a lamp, presumably to determine how long it's been out. How quickly a lamp cools to room temperature after being turned on would depend on any number of factors, including the wattage of the light bulb and how long the light was turned on in the first place. If warm, you could certainly conclude that it was lit in the not-too-distant past, but that's about it.

Aside from some of the humorous problems of this first scene in particular, this approach to the sensation of hot and cold gets us closer to how thermosensation in the skin actually works than do the scenes in which Matt senses objects near room temperature from considerable distances, and in the manner of something like sight.

We can certainly feel thermal radiation, such as from a warm fire or a cat on our lap, but what we are feeling is the transfer of heat *raising the temperature of our skin*, not the sensation of photons of infrared light striking our skin as if it were a large retina. If not even the photoreceptors in our eyes can detect photons of wavelengths below the visible spectrum, why would we expect our skin to do so? After all, a key function of our ability to sense hot and cold, from an evolutionary perspective, is to maintain a stable body temperature.

I offered a very brief introduction to thermoreceptors in chapter three, and mentioned the temperature-gated ion channel TRPV1, which is activated at temperatures above 43 °C, as well as by chili peppers! TRPV1 belongs to a larger family of receptor proteins that has at least five additional known members working in the temperature-sensing business. Two of these receptors, TRPA1 and TRPM8, detect cooling. TRPM8, which is also activated chemically by menthol and other "minty" substances, specializes in moderately cold temperatures in the interval of 8-28 °C, whereas TRPA1 is only active below 17 °C. The remaining four respond to warming with TRPV4 and TRPV3 active at moderately warm temperatures, while TRPV1 and TRPV2 respond to more extreme heat and generate a burning sensation.

As with many other sensory systems, we can get a lot of sophistication out of just a few basic components. While we shouldn't compare thermosensing to the sophistication we can squeeze out of the three differently tuned cones in the retina, we are in fact sensitive to a range of different temperatures and are able to detect minute changes. However, warm receptors are less sensitive to rapid changes in skin temperature compared to cold receptors, which also happen to be more plentiful. This makes us better equipped to notice sudden cooling than sudden warming.

. . .

This is the domain where playing around with heightened senses would be fruitful. It's an open question how much extra mileage Matt Murdock could get out of simply having a higher density of thermoreceptors, but we can certainly picture a scenario in which he is simply more sensitive to changes and differences in temperature.[4] There is a scene in the sixth episode of the first season of *Daredevil*, just before the "half a box of nails" shenanigans, where Matt notices a specific difference in temperature near the site of his adversary's bullet wound. Unlike what happens after that, this does seem like an ability a heightened sense of heat *discrimination* might afford you, especially if making such discriminations is something you've had some practice with.

Another scene where this ability is put to sensible use is in *Daredevil* #180, by Frank Miller. In this issue, Daredevil and Ben Urich go on a recon mission down in the sewer system on the suspicion that the mysteriously missing Vanessa Fisk might be down there. They happen upon a large duct that leads straight into the ground. Matt holds his hands out and notices warmer air coming from the depths below. Ben, meanwhile, feels nothing. For this to work, we have to assume that the air just above the pipe is in fact slightly warmer than the ambient temperature, but it is not unreasonable to assume that Matt's heightened "sense of heat" might be quicker to register small differences than Ben's.

READING BY TOUCH

The sensation of hot and cold obviously contributes to our touch sensations more broadly. For instance, the respective thermal properties of materials like wood and metal, help make them recognizable as such. However, it's time to turn to "touch proper," or the sensations from our skin and tissues that arise from a mechanical stimulus.

The skill we associate most strongly with Matt's heightened sense of touch is his ability to read printed text. For the first few decades of the comics, he even did this with his gloves on! It is such a staple of the Daredevil toolbox that you may be wondering why I didn't start this chapter there. Let me assure you that I have my reasons. The most important is that I regularly see fans online suggesting that what underpins Matt's print-reading ability is that he is able to distinguish lighter and darker areas of the text he is touching. I wanted to make sure that we first rule that out. However, it's worth noting that around the time Daredevil made his debut there were strange reports out of the Soviet Union – that made it into the mainstream American press! – of people who could allegedly read print by touch while blind-folded, presumably evidence of an even more extreme case of opto-dermic perception than merely sensing color.[5]

This alleged ability has been soundly debunked as a hoax, and it is to the credit of Daredevil's creators over the years that differences in color have never been the reason stated for why Matt can read print by touch. Instead, the idea behind it is that he feels the differences in *texture* between the paper and the thin layer of ink deposited on top. This naturally makes the utility of this skill highly dependent on the printing technique. Many modern printing processes leave no discernible texture on the page at all, and cannot readily be compared to the newsprint of the 1960s. This state of affairs was addressed by then *Daredevil* writer Mark Waid, in a 2012 interview:

"Unlike most comics characters, Daredevil is a character who actually gets less powerful over time, in a sense. Not physically, he still has the same powers, but think about how much of our lives we live on screen now, how much of our lives we live virtually. [...] It's a constant thing of people reading things on screen... Newsprint! He used to be able to read newsprint, but he can't anymore because even if it's not on screen, printing is not by newsprint anymore – it's offset printing which means there's no texture to it. I'm thinking about all these things constantly. How has the world changed, and how has technology left Daredevil's powers behind and made it even more difficult for him to maneuver in the world?"[6]

Print reading by touch hasn't completely gone the way of color-sensing just yet, but it is not as prominent as it used to be. For its part, the *Daredevil* television show also didn't feature any instances of Matt reading print by touch.[7]

To figure out what to make of this Daredevil staple, let us first look at the reading by touch that most people are familiar with, that is the ability of blind people to read braille.[8] In reality though, far from all blind people read braille. People who lose some or all of their sight at an older age, which is the majority of the visually impaired population, may not be willing or able to spend the considerable time and effort it takes to learn the skill well as an adult. Other tools have also come on the scene, such as text-to-speech technology and recorded audio materials.

Even among children with vision impairments, the amount of residual vision they have usually determines whether they receive any braille instruction. Sadly, many of the school-aged children who might benefit from learning to read braille are being denied this opportunity. The National Federation of the blind are even talking about a braille literacy crisis.[9]

While it lies beyond the scope of this book to go too far into the details, some of the numbers here are worth noting. According to a 2010 study, twelve percent of those who are legally blind in the United States can read braille. In the 1960s, when the *Daredevil* comic made its debut, that number was 50 percent. According to data from 2015, fewer than ten percent of blind school-aged children are currently learning braille. This despite the fact that findings consistently show that knowing braille is positively correlated with employment, level of education, and income, as well as other quality of life outcomes in blind adults.[10]

In the *Daredevil* comic, we can assume that Matt Murdock learned how to read braille between pages ten and eleven of the first issue, and he has been demonstrating this ability on a somewhat regular basis ever since. The *Daredevil* television show showcased it frequently.

Matt's relationship with braille is complicated by the fact that he can also read print by touch. The early comics would have us believe that he can do so just as easily as he can read braille, reducing the latter to simply a part of the ruse that protects his secret identity, and a burdensome one at that. To quote a caption from *Daredevil #4* (1964), where we see Matt doing research in his office: "[...] We find him rapidly scanning his braille law books, even though his super-sensitive fingers could 'read' ordinary print if he wished, merely by feeling the impression of the *ink* on the page!"

This attitude toward braille has softened considerably over the years, even though a scene from the relatively recent *Dark Reign: The List - Daredevil* (2009) #1, by Andy Diggle and Billy Tan actually has Matt decline an offer of documentation in braille that is already being provided in order to impress the Hand.

One thing braille has going for it, that printed text doesn't, is that it is perfectly adapted for the sense of touch. Before the invention and spread of braille, the scarce reading material available to the blind actually consisted of large embossed roman letters. While certainly better than nothing, this made for a slow and cumbersome reading experience. There was also no way for members of the target audience to write using this method. Louis Braille's innovation changed all that.[11]

The braille "cell," which contains combinations of one to six raised dots distributed over three rows and two columns, neatly fits the size of a human fingertip. Furthermore, the distance between the dots approaches, but does not exceed, the threshold of spatial acuity in this area. What you end up with is a code that is maximally economical while still being unambiguous and very "touch friendly."

Of the four types of touch receptors we find in the skin, two are involved in braille reading, and fine touch in general. One type is Merkel cells, which are found joining forces with a class of "*slowly* adapting somatosensory nerve fibers" (SAI). Also active are Meissner corpuscles, the encapsulated nerve endings of *"rapidly* adapting somatosensory nerve fibers" (RAI). Of the two, Merkel cells have the smallest receptive fields, and thus the greater spatial acuity.

That SAI nerve fibers are "slowly" adapting doesn't necessarily mean that they are sluggish (they are not), but that they are less preoccupied with novelty than the rapidly adapting nerve fibers. If you, for instance, place your fingertip on a small bump, the SAI fibers in the vicinity start firing excitedly and then simply keep going. The firing rate slows down after the initial burst, but it stays elevated for as long as you keep your fingertip in contact with the stimulus. When you remove it, the firing stops immediately. RAI fibers, on the other hand, are very easily bored. There is rapid firing when the stimulus is new, followed by no activity at all, and then a slight uptick when the stimulus is removed.

Due to their different proclivities, Merkel's cells and their SAI fibers specialize in the detection of edges and points, whereas RAI fibers and Meissner corpuscles specialize in lateral motion, such as when you explore the texture of a material. Braille reading involves both of these activities, since it consists of a scanning motion across a series of dots, which translates into a nice, spatially segregated on and off pattern as the fingertip reads a line of text.

You might be curious to learn how fast someone is able to read braille. The answer is complicated by the fact that the braille reading population is naturally diverse. The fastest readers are found among those who learned in childhood. Reading speeds of hundreds of words per minute are possible, but quite rare. People who learn as adults often discover that they hit a barrier of around 60 words per minute. All things being equal, though, braille reading tends to be slower on average than reading visually. This is not at all an indictment of braille as a reading method, but more a reminder of some of the natural limitations of the sense of touch, whether heightened or not. Touch allows for an impressively rapid integration of information, but is limited by certain constraints that vision is not.

There is a scene in *"Shadows in the Glass,"* episode eight of the first season of *Daredevil*, where we see Matt reading a braille document at the office. His reading speed conspicuously picks up after Foggy leaves for the day. Is this him displaying a superpower he would rather keep hidden? While I don't know what the creators intended, this need not be the case. He could simply be scanning the document, the same way people do visually, in search of specific words with little attention paid to the rest of the text. Or he could just be a very fast reader. The combination of having been blind since childhood and having a heightened sense of touch certainly has to count for something!

How might all of this apply to reading print by touch then? Well, there's a certain point along this spectrum that doesn't seem all that goofy. If you take the example of an ever so slightly embossed headline that features large letters and a noticeable texture difference, this takes us into real-world territory. What Matt does in the comics, though, is read newsprint. This challenges his sense of fine touch in at least two ways. First, it requires a *much* finer spatial acuity than anything a real human is capable of. The two-point touch discrimination threshold in the fingertip of a young adult is around 1 mm (or 1/26th of an inch). Even though a printed letter in a typical newspaper exceeds this size, the smaller features that distinguish two letters may not.

There's also the fact that printed letters may only be ever so slightly raised off the page unless printed with some embossing technique. With modern printing, the trace of ink may quite literally disappear in the noise of the structure of the paper itself, presenting us with a genuine stimulus problem. Even when there is *something* there, the "contrast" between the background texture of the page, and what is printed on top is bound to be exceedingly low.

We could, of course, have the miracle exception take care of some of this for us, and simply decide that Matt Murdock gets to have a higher density of SA1 and RA1 nerve fibers, which in turn sport more than their typical number of Merkel cells and Meissner corpuscles, and that these will start to fire at the mere hint of an indentation.

However, I think this is best addressed as an ability that exists on a spectrum where we can allow Daredevil to exceed normal human limitations while still acknowledging that dot patterns make for a *much* crisper code for the sense of touch to be presented with than small swirls of roman letters, even under the best of circumstances. There are also other things to contend with, such as the added difficulty of staying on the same line, and finding the next line easily, when the lines in question are minuscule. I get a headache just thinking about it! Glossy magazines? Forget about it.

One relatively recent example of Matt reading a sizeable amount of small printed text, as opposed to an occasional headline, was in *Daredevil* #93 (1998), by Ed Brubaker and Michael Lark, where he is seen reading Vanessa Fisk's obituary. Oddly enough, he is holding the newspaper in front of him with one hand and reading it with the other, which looks... uncomfortable. The look of this scene strikes me as oddly anachronistic, and it may be the case that successive generations of writers and artists have come to feel the same way because you don't tend to see this much anymore.

That reading this way might at the very least be a bit of a chore was suggested as early as Frank Miller's and David Mazzucchelli's classic Born Again storyline. In a scene from *Daredevil* #227, a newly awoken Matt goes to check his mail, and thinks to himself: "I read the *envelopes* with my *fingers*. The embossed ones are easy. Going by the scant impression of the *ink* on the others is a *pain*, this early in the day."

Before moving on from the various forms of reading we have seen Matt do over the years, let's also look at the blissfully short-lived ability that surfaced in the early 1990s: The ability to read computer screens by touch!

Given the perpetual demand placed on the character's creators to find ways to compensate for his blindness, I suspect that if office computers had been around in 1964, Stan Lee and the rest of the gang would have bestowed Daredevil with this ability. Alas, computers didn't enter people's homes and workplaces until decades later, at which point it became an issue to address, and the reading of computer screens by touch made its debut in *Daredevil* #298, by D.G.

Chichester and Lee Weeks, during a scene that sees Matt doing a search on a S.H.I.E.L.D. computer:

"Middle fingers find the raised ridges of the home keys, the rest falling easily into place. ...Typing out *words* I send on a microchips *quest*. 'Kingpin,' 'Murdock,' 'Fisk.' A dozen more deep within S.H.I.E.L.D.'s computers, one of them triggers a hard drive cycles... [sic] A magnetic thrumming raising the hairs on the back of my hands. An *electronic* key opening a high-tech *Pandora's box*. Characters crawl across the screen, warm *phosphors* under my *fingers* writing out an *indictment*."

We see more of the same in *Daredevil* #303, where Chichester is joined by M. C. Wyman on art duties. This time, Matt is at the office where he actually has access to screen reading technology! Still, he thinks to himself: "The machine drone of the speaking lags behind what my fingertips could 'read' in the hot-cold phosphor letterforms on screen — but the headphones remain a necessary concession to appearance and hidden dual identity."

While I can't say for sure what the output rate of synthetic speech was in early 1992 when this issue came out, it is certainly not the case that this is a slow process today. As I mentioned in chapter four, blind users of text-to-speech technology are able to understand speech at rates that might sound superhuman.

When I first read these issues many years ago, I was surprised to see Chichester's name attached to this particular artistic choice. His writing often attempted to emphasize the crudeness of the radar sense and is not someone I would have pegged as willing to minimize Matt's reliance on traditional assistive technology. Although, a mere four issues later, in *Daredevil* #306, with art by Scott McDaniel, this newfound ability fails to live up to its past glory as Daredevil loses in battle to a touchscreen:

"Information kiosk touch *screen* – warm *phosphors* under my hand. Colorful *graphics* to people with eyes that work... meaningless swirls to me. I *grope* in my darkness, tapping hard against every corner of the glass, hoping for the whirring hiss of a printer I'm finally rewarded with."

As should be clear from our earlier section on the sensing of heat and color, what is described here fails for some of the same reasons. No, the letters on screen cannot be heating his fingers in some meaningful way, much less one that maintains the resolution of the letters, and this ability has fallen completely to the wayside.

HAPTIC EXPLORATION

As we have noted earlier, the very first mention of the sense of touch in the *Daredevil* comic is not one that addresses Matt Murdock's heightened sense of fine touch. Instead, it is his weighing of a gun in his hand which gets to represent this sense.[12] This depends on the sensory receptors that gauge the amount of stretching of tendons and muscles which also underpins proprioception or the sense of our limbs and body in space. The handling of the gun though, feeling its shape and grasping it just right, additionally depends on touch receptors that are found deeper in the skin—Ruffini endings and Pacinian corpuscles, both of which are encapsulated nerve endings.

Analogous to the case of Merkel cells and Meissner corpuscles, these touch receptors are also associated with slowly and rapidly adapting neurons (SA2 and RA2, in this case). Ruffini endings are slowly adapting SA2 fibers and are sensitive to the stretching of the skin. Together with similarly slowly adapting SA1 neurons associated with the Merkel cells they mediate the feeling of sustained touch. Pacinian corpuscles, which wrap around RA2 fibers, are very sensitive to vibration. The density of SA2 and RA2 nerve fibers in the skin is much lower than is the case for SA1 and RA1 fibers. Additionally, each of the "type 2" nerve fibers is associated with a single Ruffini ending and Pacinian corpuscle respectively. The more numerous "type 1" fibers of fine touch, on the other hand, have many bushy little branches with Merkel cells and Meissner corpuscles hugging each individual branch.

The active discovery of three-dimensional objects, primarily by the hand(s) is sometimes referred to as haptic exploration. Most people probably do not devote very much time to thinking about how amazing it is that we are able to quickly and reliably identify so many different objects by touch alone, because we take it very much for granted. The same goes for the role of touch in the handling of objects. A simple act like writing something with a ballpoint pen on a piece of paper requires more than simply going through the motions. These motions depend on feedback from the pen, which in turn is in contact with the paper and the surface underneath. We adjust our grip around a sponge and a cup of coffee in response to whether it yields to the pressure we put on it, and to what degree.

Several of the properties of three-dimensional objects that are accessible through the visual modality are also available through touch. As we have established, color is *not* one such feature, but the shape of the object and several other surface characteristics are. For this reason, touch becomes an important means of exploration of the three-dimensional world for people with little to no vision.

At the same time, you don't see very much haptic exploration going on in the *Daredevil* comic. It is true that it is used to recognize or get acquainted with faces (though whether this is something blind people actually do in real life is another matter entirely). A particularly harrowing example of this comes from *Daredevil* #182, by Frank Miller where a distraught Matt sets out to exhume Elektra's body after becoming convinced that she's still alive. He thinks to himself, "I can't *see* — but my hands know your face. Even after all these years, I could never forget..."

Touch also is used to explore surfaces. However, we rarely see him use touch as a guide to the *identity* of inanimate objects. The notion seems to be that Matt has little need for getting acquainted with objects this way thanks to his radar sense. However, this line of reasoning falls into that rigid all-or-nothing thinking that affects much of the portrayal of the Daredevil character. The expectation seems to be that because Matt can sense the presence and shape of objects around him, an ability to immediately *recognize* these objects would follow automatically.

But even the more extreme interpretations of the radar sense are bound to be lacking in several features we associate with vision. Again, color is the more obvious but fine spatial resolution is another one. Compare your ability to recognize objects using your peripheral vision to what you are able to do when you look at them directly, and you will get some sense of what I mean.

My point is that we should expect touch to play a bigger role in Daredevil's life than it does in the lives of most people. Perhaps not to the extent that it would necessarily give him away in conspicuous ways when he's on patrol, but definitely to the extent that touch would have an actual role to play in his identifying and familiarizing himself with more complex physical objects that are not large in scale, or immediately recognizable by context and shape alone. An argument can also be made that touch is a more salient, psychologically satisfying experience than the radar sense because it provides a more intimate and detailed knowledge of objects. We will get back specifically to the experience of "radar-sensing" in chapter eleven.

Of course, leaning into the notion that physically exploring objects may sometimes be necessary, or at the very least rewarding, challenges some of the ideas around how Daredevil's remaining senses are expected to compensate. In the comic, we are used to seeing them do so not only in the *absence* of sight but often in the *manner* of sight. I don't suspect this is done for any kind of nefarious purposes, it is probably simply the case that most creators, and fans for that matter, are once again having a difficult time mentally divorcing themselves from the logic of vision.

It is also the case that the comic book format itself lets creators off the hook here. While skilled artists are certainly able to infuse the pages with dynamic action, the medium is still one of sequential art—a series of static two-dimensional images are presented one after the other and are tasked with providing a full story over the course of just 22 pages. Fine details that might give us insight into how Daredevil would move in a real environment will be rare by necessity. This is not the case when every segue and transition between movements can be documented by a camera.

How the *Daredevil* television show would tackle this challenge was one of the things I was most looking forward to ahead of the show's release in 2015. The format of a television show that would have a runtime of almost twelve hours for the first season alone also meant that there would be room for the kind of quieter moments and glimpses of everyday life that are rare in the comics. In this regard, I was not disappointed. I bring this up on the topic of touch because I thought this was a sense that was generally handled quite well. There is enough magic to go around, and I have covered several such scenes, but Matt's use of touch helps keep us grounded, and I would like to end this chapter by looking at a few such examples.

The scene that stood out the most to me in the first season comes from the end of episode seven, "*Stick*," where Matt has to clean up his apartment in the wake of a violent falling out with the man who trained him. Amid random debris and splintered furniture, we find Matt crouched on the floor, lightly brushing his fingers over the carpet in a way that looks remarkably like something he would have been taught to do. His fingers then find the ice cream wrapper bracelet he made for Stick as a child. While it would have been a big ask for any of his other senses to have discovered the bracelet (with the possible exception of smell), it is still worth noting that Matt does this entirely by touch.

Another noteworthy example is not a single scene, but a series of scenes taking place at the workshop of Melvin Potter. Potter is known from the comics as the gentle giant costume maker with a shaky handle on reality who every so often wreaks havoc as the violent Gladiator. In the show, Potter's primary purpose is to function as the provider of Matt's Daredevil costumes, and Matt makes several trips to his shop.

It is unclear just how much Melvin Potter knows about Matt, or is able to figure out along the way, but despite Matt's attempts at secrecy, he seems not the least bit inclined to hide his conspicuously manual inspections of the workshop, starting in episode eleven of the first season, *"The Path of the Righteous."* In the season finale, when Matt goes to pick up his first costume, he makes sure to examine it. In episode four of the second season, *"Penny and Dime,"* Matt gets a new replacement helmet, and this time he doesn't even bother opening the box that holds it all the way before sticking his hand in there. In the season finale of the second season, Matt receives a billy club as well and takes his glove off to feel it. If Matt wanted to hide from Melvin that he at the very least has *some* kind of vision problem, he is not doing a very good job of it.

There is also quite a bit of touching going on in *The Defenders* team-up show that came out between seasons two and three. One very straightforward example comes from episode two, *"Mean Right Hook,"* where Matt searches through a small box that holds his bandaids and antiseptic wipes, after getting his knuckles bloodied. This is very clearly an entirely tactile task in ways it wouldn't be if he could actually see.

A fan-favorite moment comes from episode seven, *"Fish in the Jail-house,"* when Matt wakes up in a police station after he and the other Defenders were knocked unconscious in the previous episode. He's been given an NYPD t-shirt to wear and does a manual check of his new attire. Later in the same episode, Foggy brings a bag holding his Daredevil costume, and Matt is momentarily confused by his friend's offering until he realizes what it is and sticks his hands in the bag to verify.

It may seem a bit sad that moments as obviously appropriate for touch as these ones would appear on a list like this, but I suppose we've all been a bit starved for touch over the years.

Looking at the manifestations of touch I've addressed in this chapter, expanded to include thermosensation, it seems clear that

Daredevil's creators have never shied away from applying these senses to extremes such as reading print by touch, or even sensing color at one point! But if a big part of Daredevil's shtick is that his non-visual senses help him compensate for the loss of his sight, doesn't it make sense to allow them to do just that, and to actually let Matt use his sense of touch more often than he has in the past? This would naturally entail becoming more comfortable with drawing attention to Matt's blindness, and less reliant on the radar sense to pick up the slack.

And on that final note, it's time to move on to the final third of this book. So far we've explored our known senses, and their theoretical limits. Now it's time to venture into the unknown, and to ask science – and the *Daredevil* archives! – what to make of Daredevil's famous, enigmatic radar sense.

PART THREE
RADAR SENSE

CHAPTER 9
THE PERPLEXING ORIGINS OF THE RADAR SENSE

 "But my most important new ability is in the form of a built-in radar that I seem to have developed! It enables me to walk anywhere safely, without bumping into anything! I feel a strange tingling sensation when I approach any solid obstacle, warning me which way to turn!"

Daredevil #1 (1964), by Stan Lee and Bill Everett.

We find the first description of Matt Murdock's mysterious radar sense in the origin issue of *Daredevil*. In a flashback sequence that gives us the necessary details about Matt's younger years and life-changing accident, our hero divulges the secrets of his senses and physical prowess. In the panel which accompanies the quote above, Matt can be seen casually walking through what appears to be an office space. To illustrate what is happening, the path he takes is conveyed by a dashed line, and along it are a series of "PING!" sound effects. These are meant to illustrate Matt being alerted to the presence of objects, and the choice of the "ping" sound seems reminiscent of a radar or sonar interface.[1] However, Matt's way of describing this newfound sense seems more in line with that of a blind person who has acquired sophisticated echolocation skills than with the notion of a kind of built-in piece of radar technology.

I have long suspected that, just as the totality of Daredevil's power set is inspired by the notion of sensory compensation, the radar sense, in particular, appears to borrow from the centuries-old reports of certain blind people being able to detect the presence of silent objects, some of whom we met in chapter six. However, I will admit to the assumption that the term "radar sense," as it is typically used in the comics, was coined by either Stan Lee or someone else at Marvel, and that it was chosen deliberately, perhaps to infuse it with super-human flair.

Imagine my astonishment when I came across a quote that suggested that even the word "radar" *itself* may have been borrowed from a context that linked it to "facial vision." It was during my reading of *There Plant Eyes: A Personal and Cultural History of Blindness* by M. Leona Godin that I came across the following from Dr. Jacob Twersky's autobiographical *The Sound of The Walls*:

 "Near a large object I could sense its presence and steer out of the way of it. This obstacle sense depends on noticing subtle changes in air currents and temperature, but chiefly on hearing and interpreting tiny echoes as they rebound from obstructions. It is something like the bat's method for detecting obstacles, or like radar. The sound of a footstep, in fact the slightest sound, releases the echoes, yet it does not require acute hearing, but concentrated and trained hearing, Nor does it give any sort of auditory impression – the impression is of vague sight or of slight pressure on the face. It is often called facial vision. I certainly did not suspect that it had anything to do with my ears, though I usually did not bump into things except when confused by too much noise."

I must have stared at this piece of writing for two minutes. I looked back at Godin's introduction of this passage from Twersky's memoir, paying special attention to the year it was published: *1959*. Two years before the appearance of Mole-Man in *Fantastic Four* #1 (1961), and five years before *Daredevil* first hit the stands, someone had described the so-called obstacle sense of the blind as being "like radar." There is even mention of air currents (and temperature), as well as the confusion caused by too much noise. After writing thousands of words on Daredevil's radar sense over the course of fifteen years, I had stumbled upon what may have been a smoking gun. I hadn't even considered that I should have been looking for one.

It's impossible to know whether the use of "radar" to describe Daredevil's object-sensing abilities drew directly from Twersky's work, or was the result of the specific term having been introduced into the broader culture earlier. After all, the misnomer "bat radar" is still commonly encountered. It may even have been somewhat common to describe this particular ability of some blind humans as "like radar," long before Daredevil came on the scene. Whatever the case may be, familiarizing myself with Twersky's work has further convinced me that it is highly likely that the original use of the term "radar," in the context of the *Daredevil* comic, was intended as a metaphor. As we shall see, events in the comic itself support this notion.

HUMBLE BEGINNINGS

What is perhaps most striking about the depiction of the radar sense in the first few issues of *Daredevil* is how generally useless it is. We began this chapter with a look at its first appearance and might walk away from that encounter thinking that it is basically an ability to detect and steer clear of obstacles. For the first two issues, there is little sense that Daredevil is detecting forms of any kind. In *Daredevil* #2, Matt does indeed think to himself that "I can visualize the shape of the Baxter Building by the sound of the air currents hitting it... Just as easily as if I could *see* it!" However, nearby people are recognized by their footsteps and breathing, flags identified by their flapping in the wind, and objects haphazardly found through a combination of touch and sheer luck.

In *Daredevil* #3, there is some hint that Matt is able to sense objects smaller than the Baxter Building for more than simple avoidance purposes, as he thinks to himself: "Luckily, my built-in radar sense tells me how far I am from a solid obstacle... So I always know when to dodge and when to grasp!" However, spatial arrangements on a larger scale are deduced through a combination of sound cues and quick thinking. When Daredevil and Karen are held captive by the Owl, Daredevil thinks to himself: "I'm in some sort of large *cage*! Judging by the slight air movement above me, it's dangling exactly eight and a half feet from the ceiling... And from the sound of the Owl's voice, it's ten feet from the floor! Estimating the time it takes an *echo* to bounce back from the walls, it's a huge room, roughly 325 feet in diameter! That tells me all I need to know... for *now*!"

Let us take a moment to consider what is going on here. Over the course of the first few issues, Daredevil has sensed some remarkable things, including those that lie outside the typical set of human senses, such as electricity, and evil auras. His senses have been described as razor-sharp. But there's an odd gap in his perceptions that remains unexplained.

In *Daredevil* #4, when Matt and Karen climb the steps of the courthouse, Karen comments on how quickly Matt moves up the stairs. He thinks, "I'm getting careless! I mustn't let on that my extra-keen senses take the place of my eyesight!" But the problem is that his extra-keen senses, *as they have been described thus far*, don't come close to explaining how Daredevil is able to accomplish *any* of his most advanced physical feats.

This is also a contributing factor to what makes his space flight in *Daredevil* #2 so utterly ridiculous. The gap between the job to be done and the sensory tools at Daredevil's disposal is simply too vast. In all his encounters with people and objects thus far, Matt appears to be almost entirely reliant on their being active *sources* of sound; producers of heartbeats, footsteps, breath sounds, and rustling in the wind. But the so-called "obstacle sense" of the blind (aka "facial vision," aka echolocation) reveals to us that it is, in fact, possible to detect both the location and shape of otherwise silent objects. In grossly underselling Daredevil's ability to sense things in this manner, the creative team has to oversell what is possible to achieve *without* it.

Given the odd state of the radar sense, Daredevil was bound for a power upgrade sooner rather than later. Stan Lee and Wally Wood take baby steps in this direction in *Daredevil #5*, which sees Matt face off against the Matador. The issue starts out modestly enough. After having heard of the costumed villain on the news, Daredevil searches for him by ear and finally identifies him by the fluttering of his cape.

Later, Matt encounters the Matador again at a costume party, and we get to see his radar sense put to use as he spots the villain opening a safe. "My radar sense indicates he's standing against the wall... his back towards the crowd! I hear him turning a dial... Softly... Cautiously..." That Matt can provide details about the Matador's posture is a clear indication that he perceives the man's physical shape and not just his location.

During the commotion that ensues when the Matador makes himself known to the hapless party goers, Matt is able to change to Daredevil, but what awaits him appears to be too much to handle. The presence of a large crowd confuses him ("It's hard to separate one sound from the other!") and so does the sound of the Matador's cape. A distraught Daredevil thinks, "His rustling cape! It vibrates the air around him, fogging my radar sense!" As much as he tries, Matt eventually has to accept defeat. "He's thrown his cape over my head! Now my senses are *all* dulled! For the first time, I feel the way an *ordinary* sightless man might feel in a battle!"

To say that this is *not* one of Daredevil's proudest moments would be an understatement. However, interspersed with scenes from this ill-fated battle are a couple of panels meant to inform the reader of why Daredevil has failed so spectacularly. The first of two panels shows a schematic drawing of Daredevil using his radar sense under ideal circumstances. Arcs of lines that appear to emanate from his head are shown bouncing back from the outline of a box-shaped object. A large "PING!" highlights the spot where these "waves" touch the object. In the second panel, intended to illustrate how Daredevil is affected by too much noise and commotion, these patterns of concentric lines are overlaid with others coming from all around. There is no "ping" and several points of impact appear in places where they shouldn't be.

The accompanying captions read: "Explanation of Daredevil's radar sense: Normally, my radar sense goes out, hits objects around me, and bounces back, giving me a mental picture of my surroundings! But when there is too much movement and confusion all about me, the 'picture' which comes back is garbled and distorted!"

This is the first attempt at any kind of real explanation of the radar sense. This is also the first time we see a hint of what will appear in full a few issues down the line: "radar rings" around Daredevil's head. What are they? Going only by what we know so far, it seems more likely that we are dealing with a sound-based signal than anything else. Why? Because Daredevil's ability to make use of the signal is so easily disrupted by other sounds, as well as "vibrating air." An electromagnetic radar signal would not be the least bit affected by either of these things.

Another clue is found in the way additional "radar rings" coming from sources other than Daredevil's head are depicted in the second panel. "Too much movement and confusion," wouldn't generate radio waves. Moving an object might make a sound, and even stationary objects create echoes, but this is a phenomenon distinct from the reflection of electromagnetic waves. No solid evidence has been introduced to suggest that the radar sense is meant to be literal *radar*, aside from its evocative name.

Radar rings make their first *in-story* appearance in *Daredevil* #8 where they are represented by concentric circles that cover the background behind Daredevil as he's trying to steer a runaway car safely down the street. This is also the first issue to feature a panel that attempts to depict Matt's radar-sensed *point of view*. What we are presented with is a black and white shot of the street in front of Daredevil (still driving the car) that is quite limited in detail. A silhouetted figure filled in with a solid yellow color informs us of the presence of a man about to step onto the sidewalk. Stan and Wally try to explain to the readers what is happening:

> "Unable to stop the speeding car, the sightless wonder steers for dear life, guided by his uncanny audio-sensory power! [...] With iron nerve, with steady hands, the man without fear guides the doomed vehicle at breakneck speed through the city streets, guided by his radar sense the same as a jetliner flying through a heavy fog...! Though Daredevil 'sees' images rather than actual sights, so accurate is his sensory perception that he steers the onrushing car with the skill and precision of a master driver!"

Let us try to unpack some of this. By invoking the concept of the jetliner, in writing as well as by including the image of an airplane framed by concentric rings in the adjacent panel, a direct comparison is made between flight radar and Daredevil's abilities. However, we have also been told, just a few panels earlier, that he is being guided by his uncanny "audio-sensory power."

I don't think anyone attached to this comic book much cares that real flight radar isn't sound-based. The point is merely to bestow our hero with some kind of credible object-sensing ability. Still, you sense a certain resistance to equating this ability with normal eyesight. I can find no other reason for making the rather nonsensical distinction between "images" on the one hand, and "actual sights" on the other.

Matt's powers continue to oscillate between two extremes for the rest of the issue. Interestingly, we also see his radar rings – thinner this time – superimposed on a radar perspective background that shows the outline of a person through the wall, which is another first.

Matt ponders the identity of this mystery man, "He's moving closer! Ordinary walls are no obstacle at all to my hyper-acute extra-sensory 'radar'! I'd better learn who it is, although I've got a good *hunch*...!" Is the radar *"extra-sensory"* now?

Having taken these steps in the direction of a more sight-like radar sense, our creative team is not holding back in *Daredevil* #9. Matt encounters Foggy putting in their office and thinks, "The ol' son-of-a-gun! I can sense him holding a golf club clear across the room!"

Later, on his flight to the fictional country of Lichtenbad, Matt is able to sense the scene below: "Can this be Lichtenbad? I sense a walled city... Like a medieval fortress! Or, perhaps it's even more like a huge, present-day prison!" Inside the fortress, the full X-ray powers of the radar are on display: "My radar senses can 'see' through the floor below me as though it's made of glass! This castle isn't as old-fashioned as it seems! It's filled with complex electronic equipment... and enough power to blow up half a continent!"

At this point, I'd like to direct the reader's attention to both the use of the plural in "radar senses," as well as to the quotation marks around "radar" in Matt's internal monologue from the previous issue, *Daredevil* #8. In fact, keep your eyes peeled for these kinds of inconsistencies whenever you get the chance to revisit the older comics. You will see them everywhere!

Daredevil #10 marks Wally Wood's final issue as the regular artist, and one on which he is credited as the scripter as well as the penciller, with layouts provided by Bob Powell.[2] It is an entertaining crime mystery of an issue featuring the mysterious Organizer and the Ani-Men, a team of villains made up of Cat-Man, Bird-Man, Frog-Man, and Ape-Man. Parts of the story are quite a mess though. For one, it features a panel of Matt tuning into a radio broadcast: "What's that?! My radar hearing has picked up a radio message! It's being transmitted from this yacht!" Of course, what is likely happening here is not Matt deciphering a radio signal, but hearing what is said by the speakers on both ends, but the fact that this isn't crystal clear to the reader is... not great. Add to this that he is also able to sense "radio impulses being transmitted" later in the issue, and readers can be forgiven for scratching their heads.

The story continues in *Daredevil* #11, with Stan Lee back on writing duties. On the radar side of things, a few interesting things happen. When Matt "spots" a man ransacking their office through the building wall, the accompanying caption reads: "But, just then, the sightless attorney's uncanny natural *radar instinct* takes over..." *Instinct*, huh? *Daredevil* #11 also features an odd panel in which Daredevil appears to be using the radar sense almost a form of remote viewing or psychic ability. In the chaos after a final showdown with the Ani-Men, Daredevil needs to find the people at the top of the conspiracy. We see our hero pictured against a full radar ring background, hand stretched out in front of him as if he's about to cast a spell or exorcize a demon. The caption reads: "Then, once away from the others, the crimson crimefighter again puts his uncanny *radar sense* into play...!" Matt thinks to himself, "I've got to concentrate... To separate the thousands of sounds and mental images I'm receiving... until I find the one I seek! And there it is! In the next room!" Would this be an example of that previously mentioned "extra-sensory radar"?

Daredevil #12 is the first issue that sees Daredevil temporarily lose his heightened senses. In a guns and explosives fight against Ka-Zar villain the Plunderer, Daredevil finds that "My radar sense... it's gone! I can't visualize what's happening around me!!! [...] Now I really *am* a blind man!"

The unconscious Daredevil is carried by Ka-Zar back to his cave. Ka-Zar manages to find some "ju ju berries" to treat Daredevil, and as the berries take effect, Matt senses the transformation. "My head! — That tingling feeling — as though my powers are returning!" Over the space of a couple of pages, Daredevil's condition improves, and we are treated to a brand new kind of radar perspective panel, one in which red lines on a black background etch out a very detailed sketch of Ka-Zar, with whom Daredevil is now being held prisoner. "My radar sense is working perfectly again! I can 'see' Ka-Zar rushing to attack me as clearly as if I were truly sighted!"

SILVER AGE SHENANIGANS

We are very much still at a stage where there is a lot of experimenting going on. Stan Lee and the rotating roster of artists are trying out new things in the senses department, and we get to see it play out in real time. Lee remained as the writer on the book through *Daredevil* #50, before handing the reigns over to Roy Thomas who wrote Daredevil for the next twenty issues. Fan-favorite artist Gene Colan came on as the regular penciler with *Daredevil* #20, while the art duties for issues 12 through 19 fell primarily to John Romita Sr.

Detailing what happened on the radar front during what remains of *Daredevil's* Silver Age to the same degree as I chose to do for the first dozen issues of the comic would be excessive. However, there are some interesting twists and turns (and hilarious storylines) that deserve some amount of attention.

The confusing ways of describing the radar sense continue from the first few issues. Matt again senses people, specifically Foggy and Karen, through walls in issues #18 and #19, but rather than becoming a radar sense staple, this ability is used sparingly during the remainder of Stan Lee's run as the writer. In *Daredevil* #21, for instance, Daredevil comes up against an iron door, and explicitly acknowledges that "...even my radar sense can't tell me what's *behind* the door!" Perhaps the realization that there are compelling narrative reasons to keep some obstacles in play is what prompted the creators to tone this one down a bit going forward.

On the more modest end of things, when Matt senses Foggy being dangled out the window(!) by Spider-Man in *Daredevil* #17, the scene reads like a throwback to the first few issues: "That sudden shift of air! Someone at the window! He's dangling Foggy outside!" In *Daredevil* #19, our title character even collides with a glass table mid-fight. "Blast it! I didn't sense that table in time! I dropped him!" This is certainly a first, and a good illustration of how the radar sense varies depending on the specific requirements of the story.

The radar also continues to be identified with hearing (or air currents), on the one hand, and as a separate mode of perception on the other. In *Daredevil* #21, Daredevil is being held captive in one of the Owl's signature cages, and notices "That sudden rush of *air* below me — it's a deep *pit*!! Even *worse*!! From the way *no sound* echoes back, it's virtually a *bottomless* pit!" Later in the same issue, he gratefully acknowledges that "...a super sense of *hearing*, and of *touch*, plus a built-in *radar sense* can compensate quite a bit!"

A very entertaining sequence of events takes place in *Daredevil* #30 that has the added benefit of shedding some ever-needed light on the state of the radar sense at that particular point in time. Dedicated *Daredevil* fans will know that following Spider-Man's snooping around the law offices of Nelson & Murdock, and accusing Matt Murdock of being Daredevil, Matt is forced to invent a twin brother in *Daredevil* #25. The fictional twin, Mike Murdock, first appears in the following issue and appears as a regular until his "demise" in *Daredevil* #42.

And so, in *Daredevil* #30 Matt finds himself in the ridiculous situation of having to pretend to be his own brother being Daredevil, pretending to be Thor. This in an attempt to lure out Thor villains Cobra and Mister Hyde. However, this layering of personas and costumes makes things particularly complicated for our blind hero. Out on the town as "Thor," Matt notes that it is "Too bad I have to wear this nutty cape! By fluttering this way, it muffles some of the sound vibrations that guide me!" So, "sound vibrations" it is!

Another absurd storyline, featuring none other than Doctor Doom, begins in *Daredevil* #37, continues in *Daredevil* #38, and ends in *Fantastic Four* #73. For reasons that are not entirely clear, Doctor Doom decides that switching bodies with Daredevil makes for a rock-solid step in his plan to get back at the Fantastic Four. The last page of *Daredevil* #37 sees the actual body switch take place, and as readers, we can only marvel at the fact that Matt, suddenly occupying the body of Victor von Doom, is more surprised to be seeing his hands "encased in metal gloves!" than to actually be *seeing* them. This slight goof seems to have occurred to Stan who makes sure to have Matt at least make note of it at the start of the next issue.

However, it is Doom himself who is our main interest here. After all, having someone else occupy Daredevil's body has the potential to give the reader some amazing insight into how the world appears from his vantage point. What do we get? Well, Doom (a supposed genius) cannot quite figure out what is going on. "Strange... I seem to *sense* things rather than *see* them! The world appears *different* to me than before! My vision is *clouded*... yet *sharpened*, at the same time! Perhaps it is due to the opaque *eye filters* on this costume!"

But wait, it gets better. Next, Doom thinks to himself, "I suspect I have found the *key* to Daredevil's unique *ability*! By obscuring his normal *vision*, he has found a way to actually sharpen the use of his own *senses*!" The final word in that sentence has an asterisk attached that takes us to an editorial caption, signed "Secretive Stan," which reads: "Although he's on the right track, even Doc Doom doesn't suspect the entire, fantastic truth... namely, that DD is totally *blind*!"

A few things come to mind when reading Doom's account of his experience in Daredevil's body. The first is that Doom's reasoning about the surprising net benefits of obscuring one's eyesight reminds us of points raised in chapter one. This is the classic "sight as distraction" trope. The second thing of interest is that Doom experiences "radar-sensing" to be close enough experientially to normal vision to be even briefly confused for the latter.

This seems like a stretch, though perhaps not quite as extreme as one might think. We already know from the research record on human echolocation that sensations coming from one sense modality can be perceived to be coming from another. This also lines up well with the research we covered on sensory substitution. However, there are also aspects of the visual world that cannot be perceived by *any* of the senses Daredevil possesses, and these should be hard to miss.

We skip ahead a few issues, to *Daredevil* #46, for another scene that is interesting from a sensory perspective. At the start of the issue, Matt is in a real pinch, having been brought in on false murder charges after the machinations of the Jester in the previous issue. Waking up in the hospital wing of a full-security prison, still in his Daredevil costume, he needs to come up with a way to escape. He finds a white coat by the scent of fresh detergent (as mentioned in chapter seven), removes his mask and does his best to impersonate a doctor. After fooling a couple of guards looking for the escaped Daredevil, and throwing them off his scent, Matt sets off down the hall.

"*Okay*, radar sense — don't fail me *now*! I have to look like I *know* where I'm going! Too many *guards* down the hall — may still be *watching* me! I'd better make sure I *find* the med supply room! Behind this *door* — all sorts of *pharmaceutical* odors! I've got to *chance* it!" Opening the door, Matt notices that "It's *empty* — but judging by the *echo* of the sound of the *door* opening — there's *another* room behind it! Having a built-in *sonar sense* can be mighty *handy* at a time like this!"

A door with glass panels leading to a separate room is visible on Matt's left – the reader's right – as he enters the first room, and it is presumably the way the different surfaces differentially reflect the sound that informs him of the presence of the second door. This is an unusually well-written scene overall that even includes the oft-forgotten sense of smell. It also seems to make the case that our hero is relying on good old echolocation, with the radar sense suddenly becoming a "sonar sense."

This scene also serves to remind us that even nearly fifty issues into the history of the *Daredevil* comic, we still don't have a clear and unambiguous definition of what exactly the radar sense is, or how it works. Many different explanations have subtly been put forth, but they vary between, or even within, issues and even contradict each other.

When the writing duties eventually passed from Stan Lee to writer Roy Thomas, the latter had the benefit of still being paired with the book's longtime artist Gene Colan. Thomas's run spanned twenty issues, although *Daredevil* #53 was a near-verbatim retelling of *Daredevil* #1, and *Daredevil* #69 was written by guest writer Gary Friedrich. Among the run's memorable highlights is the storyline from *Daredevil* #56 through #58, during which Matt reveals to Karen that he is secretly Daredevil! (Matt chooses to make this shocking confession immediately following the funeral of Karen's father Paxton Page, making it yet another in a long line of questionable life choices.)

What about the radar sense? While Stan Lee remained as the editor, there was always the chance that Thomas, as the new writer, might pick a strategy and stick to it. Not so. The radar sense continues to oscillate between modest and extreme, and we even see the return of the "X-ray" radar in Thomas's first issue, *Daredevil* #51.

In a story that continues from Stan Lee's last, both with art by Barry Smith, we see Daredevil fight an evil robot called the "Plastoid." In one panel, we are shown a detailed radar perspective shot, complete with superimposed concentric rings of the robot and all of its otherwise hidden components. Unlike the skeptical police on site, Daredevil is certain that the Plastoid is about to explode. "This is hardly the time to spill the beans about my radar sense! I've located the destruct mechanism... and something else! Something that tells me — time's running out!" DD ends up saving the day by striking the robot's chest plate hard enough with his billy club to somehow jam the trigger.

The actual *nature* of the radar sense also remains somewhat unclear throughout this run. On the one hand, we have the scene from *Daredevil* #63, in which the radar sense appears to be very much distinct from Matt's sense of hearing. In the middle of fighting the Gladiator, Daredevil thinks to himself, "*Wait! Now* I sense him! The *siren* drowns out his heartbeat — but my *radar sense* tells me that he's coming at me —*fast!*" Other scenes in other issues feature similarly clear distinctions. However, we also commonly encounter "radar rings" being used in contexts that showcase Daredevil's sense of hearing.[3]

On this particular topic, it is also worth taking a look at the way Matt describes his senses to Karen. From *Daredevil* #58: "As Matt, I told you once about the childhood accident that blinded me! That story was true, but not the whole truth! For, in some mysterious way, the same mishap that robbed me of my sight... Amazingly sharpened my remaining senses, to far beyond those of other men... enabling me to avoid disasters, and to perform athletic feats that few people even dream of! Taste... touch... smell... hearing... All my senses were heightened!"

Matt makes no mention of the radar sense just yet. However, when Karen waves her hand in his face, he elaborates further. "You can save your energy, beautiful! Maybe I can't see, but I've got a radar sense that tells me you're waving your hand... testing me!" Just as in the very first issue, the radar sense comes last. Is it an afterthought or something that is meant to follow from what has already been said?

Before moving on, we should pause to consider one pitfall of trying to analyze Silver Age comics from the perspective of a superhero's power set. The problem is that when these early stories get weird, they get *really* weird. Logic and consistency frequently go out the window, and as a reader looking back on this era, you easily fall into the trap of remembering the more absurd moments out of proportion to their prevalence. It is only in studying these early issues in great detail, with as much attention paid to the less spectacular events as to the obvious outliers, that you get a real sense of what the standard use of Daredevil's senses was at this time.

The pattern that emerges is that of a fairly modest interpretation of what Matt is able to do with his powers. While we don't know definitively what the radar is supposed to be, its practical uses are easier to follow. In fact, the outliers strike us as bizarre in large part because of how sharply they contrast with this underlying pattern.

To take one example of how this plays out, let us look at *Daredevil* #24, by Stan Lee and Gene Colan. After being teleported by the Masked Marauder straight from New York to an unknown location in Europe in the previous issue, Daredevil finds himself in a pickle with a gun aimed in his direction. The first few pages are spent in battle, and the way he deduces that he is "confronted by *armed men*..." is by counting "Six pulse beats in front of me!," and hearing "The sound of *triggers* being cocked!"

This kind of scenario, where Matt's ability to sense the shape of people and objects is called upon only *after* other sources of information have been tapped – and sometimes not at all – has been common throughout much of Daredevil history. I have taken to calling the phenomenon CAR, or "conspicuously absent radar," and I'll give it a closer look in chapter eleven.

Getting back to the story of *Daredevil* #24, we learn that Daredevil is able to escape the armed gunmen and then happens upon an airplane on a beach below him. "There's a *clearing* below me... with a strong scent of *jet plane fuel!* That means just *one* thing... I've found my ticket *home!* Only *one* heartbeat in the area! A solitary *guard* is near the plane! But... not for *long!*"

Daredevil apparently figures out that there is a plane in front of him *by the scent of jet fuel*. This sets us up for a narrative whiplash when Daredevil next gets behind the controls of the plane. As he takes off, after punching the guard, he notes to himself: "I'm in *luck!* There's enough fuel to reach *England!* And, from *there*, I'll hop on a transatlantic jet!"

Of course, Stan Lee *knows* that this is a far-fetched use of Daredevil's powers. Hoping to preempt an onslaught of fan outrage, Stan addresses the readers at the bottom of the page: "To save you the trouble of writing scathing *letters* to us, we'll explain here and now *how* the sightless D.D. can pilot a plane! He feels the vibrations of the needles and dials within the instrument panel, and his own natural *radar sense* takes care of the rest!"

Of course, this kind of explanation didn't make sense in the infamous *Daredevil* #2, and it doesn't make sense here either. How does Matt know how much fuel it takes to get to England when he doesn't even know where he is? What direction should he be going, and where will he find an airport? And how on Earth does the radar sense, treated as an afterthought moments earlier, suddenly become powerful enough to track the ground thousands of feet below?

These are the kinds of scenes that tend to stick in the reader's mind, and for good reason. But they are not representative of any kind of standard take on the character, even at this moment in *Daredevil* history, and even with the considerable confusion surrounding what makes "radar-sensing" possible in the first place.

. . .

Before leaving the Silver Age behind, and adopting a more thematic approach to the radar sense, let us take stock of where we are and what these themes we are going to investigate could possibly be.

It is not an exaggeration to say that the state of the radar sense has been a bit of a mess so far. Certain patterns and broad strokes have emerged when it comes to both the functional utility and experiential nature of this mysterious ability – more on that in chapter eleven – but its physical nature is far from settled. Instead, what we are left with going into the later runs of the comic are a handful of different explanations that successive writers have latched on to and managed to standardize – to a degree.

What are they? The first distinction we need to make is whether the radar sense is, in fact, a real *physical* sense or something "extra-sensory." While most interpretations align with the former, we did see hints of the latter in a couple of the issues we have covered and the idea has occasionally resurfaced.

Looking instead at what appears to be the majority take, that is the radar sense as something physical and "of the body," we are still left with a few different explanations. One is that the radar sense is exactly that which I suspect Stan Lee, Bill Everett, and others originally imagined it to be – a heightened-senses version of the ability exhibited by some blind people in real life. In this context, the term "radar" would be understood as a metaphor for echolocation. Even when the sense is portrayed as being separate from hearing, it may still be the natural "obstacle sense" that the writer has in mind.

Not surprisingly, the comics have been as confused about what actually supports this ability as the research record was prior to the experiments at Cornell, and so we see references to "radar senses" (plural), "radar hearing" and the like. There are also frequent mentions of air currents that also tie in well with the early understanding of "facial vision" as a phenomenon. As we saw with Matt's escape from the prison ward in *Daredevil* #46, there have also been scenes that take us very close to an understanding of the "radar" as echolocation *specifically*.

Sitting somewhere between the radar as a metaphor to describe the combined efforts of Matt's remaining senses (with a natural focus on hearing) and a bona fide sixth sense, we find the idea of the radar sense as a physical sense, but one that is a distinct sensory modality with no equivalent in nature. This interpretation has Daredevil emit physical, albeit mysterious waves and then registering what echoes back. This is the interpretation of the radar sense that I would describe as the most common when considering Daredevil history as a whole. However, as I have argued previously, it does not appear to have been the *original* understanding. It also takes considerable time for it to emerge as the "standard," and this is a topic I will return to in the next chapter. In the comics we have covered so far, we have yet to see it fully formed. It has certainly been hinted at, but so far seems more like an also-ran in the confused battle for radar domination.

I suspect some of you are already getting ready to object at this point. Wouldn't the frequent appearance of visible radar rings in the *Daredevil* comic, going all the way back to *Daredevil* #8, be a clear indication that the idea of literal waves flowing from Matt's head has always been common? Well, yes and no.

Aside from the generally confusing writing on this matter, radar rings sometimes appear out of context. And even though we see frequent references to the radar as a separate sense, the term "radar" might still be understood as a metaphor. For all we know, it could actually be sonar. In fact, this appears to be the suggestion made even much later by David Michelinie, the guest-writer of *Daredevil* #167. In his words: "Daredevil has an uncanny *radar-sense*. Like a bat, he emits probing high-frequency waves. Waves which break against any solid object, and breaking, send back signals audible only to Daredevil. From these signals, his brain instantly forms silhouette images of everything around him. In this manner, he 'sees' in every direction!"

We should also remind ourselves that visual metaphors are common in comic books, and function as narrative devices. An exclamation point can be added to a panel to illustrate surprise on behalf of a character. It doesn't mean that a physical object in the shape of an exclamation point appeared in the story. In the realm of Marvel comics, readers will also be familiar with the visual shortcut that frequently accompanies the tingling of Peter Parker's spider-sense. It is typically drawn as squiggly lines around Peter's head, but occasionally readers will also see a Spider-Man mask appear to cover half of his face. None of this is to be taken literally, but serves only as a way to tell the reader that his spider-sense is altering him to some danger.

A similar case could easily be made for visual depictions of Daredevil's radar sense. There is no doubt that the waves (lines) around Matt's head have come to be understood as literal by some (perhaps most) creators, but I am far from convinced that they started out that way, or that this was the consistent view during *Daredevil's* Silver Age.

Before we move on, there is one rather humorous exception to my conclusions above that needs a special mention, and it is one that is so oddly specific that it makes me chuckle to think about it. The one scenario in which Matt's "radar sense" absolutely becomes a literal *radar* sense is whenever he steps onboard an airplane. This appears to turn this otherwise modest sense of objects in Daredevil's immediate vicinity into a near-magical ability to sense the ground *thousands of feet below.*

We have already encountered this ability during Matt's trip to Lichtenbad in *Daredevil* #9, and his escape to London in *Daredevil* #24. Amazingly, the ability exhibits an impressive amount of staying power, and survives into the so-called "Bronze Age," of comics.[4] In *Daredevil* #85, by Gerry Conway and Gene Colan, Matt and Natasha Romanova (the Black Widow) are on their way back from Europe on a Boeing 747 when the plane is hijacked by the Gladiator and his team. After switching to Daredevil and subduing the attackers, Matt then helps to guide the plane to safety using his radar sense! We even get to see Daredevil pawing around at the control panel in the cockpit as if he knows what he's doing.

In *Daredevil* #100, by Steve Gerber and Gene Colan, we see another example of Matt's supposed skill as an aviator. The issue opens with Daredevil sitting at the helm of an Avengers' Quinjet, drawn as if he's looking out the window. While there is no actual mention of the radar sense, the caption informs us that, "His hyper-sensitive fingertips read the dials and meters his eyes cannot see. And he knows that he is... home?"

CHAPTER 10
MAKING SENSE OF THE RADAR SENSE

 "Now he had a sensation of waves creeping out from somewhere behind his eyes. The waves lapped against the sliding door that opened onto the balcony and carried back the flat hardness of plain glass: he had forgotten to draw the blinds last night. He felt warmth and knew that he would not need a heavy jacket today, but the smell of ozone in the air meant he should carry a raincoat for later. He stood in his darkness and let the sunlight warm him for a moment."

Blind Justice, by Kyle Christopher (Featured in *Stan Lee presents The Marvel Superheroes*, Marvel Novel Series #9)

The quote above comes from a prose Daredevil story written by Marty Pasko (pseudonymously) in 1979.[1] I was first alerted to its existence when writer Mark Waid spoke highly of it in a couple of different interviews during his run(s) on *Daredevil*. Knowing that this book was out there quickly prompted me to find a time-worn used copy for myself and dig in. I agree with Waid's assessment that Pasko had an amazing knack for writing Daredevil's senses. The story provides an intimate and visceral experience from Matt Murdock's point of "view," and the fact that its format demands that all necessary information be relayed solely by the text on the page makes it even more intriguing.

As is evident from this brief introductory snippet, Pasko also had a particular interpretation of the radar sense in mind when he wrote it: That of literal waves flowing from Matt's head. Even though this particular interpretation of the radar sense (a separate sense that depends on Matt being an active source of the signal) had been hinted at – albeit inconsistently – for many years, we have to look at Frank Miller's first run as the writer of Daredevil, which began in late 1980, for similarly bold statements about literal waves flowing from Matt's head.

Given the prominence of Frank Miller's work, it is perhaps not surprising that his run is able to set something of a standard for later writers to latch on to, particularly for the remainder of the first volume of *Daredevil*. (There are also some notable exceptions we'll return to which rather ironically includes later versions of Miller himself.) However, during most of the "Bronze Age" runs which preceded Miller's, the radar continued to be in a state of flux, reminiscent of the even earlier stories we looked at in the previous chapter.

The first writer of that era, Gerry Conway, typically used the plural "radar senses," and even had Matt explicitly echolocate in *Daredevil* #87. However, in a later stint as the guest writer of *Daredevil* #118, he described the sense as "radar-vision." Then-editor Roy Thomas even provided a caption which explained it as "[t]he radioactively induced sixth sense that helps our sightless swashbuckler 'see.'" Steve Gerber, on the other hand treated the radar as a separate sense in his writing and referred to it in the singular, though he made a frequent point of it being sensitive to noise. When Marv Wolfman came along, there was a return to a somewhat consistent plural "radar senses."

However, *Daredevil* #142, by Wolfman and with art by Bob Brown, provides us with an excellent example of the still-fluid nature of the radar sense. At the beginning of the issue, Daredevil regains consciousness only to find himself in a spectacular pickle courtesy of the recently introduced Bullseye. "My radar senses tell me I'm tied to an arrow, and I don't need anything to inform me I've just been shot from the bow." He adds that it "Doesn't take much to bounce my radar off whatever I'm rushing at. That's the New Jersey palisades I'm going to splatter against in less than thirty seconds!" Here, we see both the plural "radar senses," as well as the singular "radar sense," which can apparently be actively bounced against something. As if we weren't already confused, Matt also notes later that "My radar hearing... it's picking something up from that rooftop —" This all happens *in the same issue*!

The radar sense went back to being mentioned in the singular with Roger McKenzie as the writer, beginning with *Daredevil* #151. It was during this time that the young Frank Miller joined the book as the artist with *Daredevil* #158 before taking over writing duties as well with *Daredevil* #168. As you may recall from the end of the previous chapter, *Daredevil* #167 was written by David Michelinie (with art by Frank Miller) who stated explicitly that the waves Daredevil emits are *audible*, and that the mechanism is like that of a bat.

THE RADAR ACCORDING TO FRANK MILLER

Before going to the comic book record, let's begin with a very interesting look at what Frank Miller himself had to say about Daredevil's senses in an interview with Dwight Decker for issue #70 of *The Comics Journal* which is dated to January 1982, about a year into Miller's solo run. Decker says, "I'd like to discuss the extent of Daredevil's powers. In one of the issues scripted by Roger McKenzie, Matt Murdock was shown hearing the Hulk's heartbeat from four blocks away." Miller's response:

> "That's pretty extreme. When I started writing the book, I sat down and defined for myself exactly the parameters of his powers. I think he has the potential of being very believable. The way to make him credible is

to have his powers be extraordinary enough to be exciting, but not on par with Superman. One of the things
I've done recently is revamp that radar sense of his to
make the images he receives less distinct. I want it to
resemble the "proximity" sense that some martial
artists claim to have, where they can detect movement,
form and location, but they don't get pictures. I'm not
fond of the radar sense in the first place. I haven't given
myself the opportunity to explore it very much."

Considering the steps Miller took to push Daredevil's hearing to
new extremes (see chapter five), I find some of what he says here
surprising. That includes his claim to not be fond of the radar sense.
The character Stick had just been introduced to the comic and tasked
with literally restoring Daredevil's lost radar sense, and Miller's first
few issues appeared to do more to elevate the status of the radar as its
own distinct sense, complete with active waves (of nondescript origin)
than to challenge it.

One legendary scene from *Daredevil* #169, Miller's second issue as
the writer, sees Daredevil fight Bullseye in the subway where the
former struggles against the onslaught of noise from the train: "Even
his radar sense fails him. As ever, the waves flow from his brain,
probing the world about him, but the descriptive signals return to a
brain that is stunned, confused..." Matt thinks, "My hands... shak-
ing... No... not my hands... It's the tracks... yes... the tracks are vibrating... Can only mean... there's another train... another train... coming
this way... Still can't hear... radar gone... can't tell how close train is..."

This scene provides what is at this point an unusually literal take
on the radar sense. Just as in Marty Pasko's story, we are told of
flowing waves. We also learn that the commotion of the subway
wreaks havoc on Daredevil's ability to function not because it is
disruptive to the signal itself, but because it affects his ability to
process it. This is not the first time intense sensory stimuli have made
it hard for Matt to concentrate – Miller's predecessor Roger McKenzie
often used physical pain to challenge Daredevil this way – but it is

quite a departure from some earlier interpretations, flimsy as they may have been.

For an example of something that *does* interfere specifically with the radar signal during Miller's run, we can look instead to *Daredevil* #178. In a rather humorous issue that sees Foggy contract the Heroes for Hire – Luke Cage and Iron Fist – to serve as Matt's bodyguards, the unwilling recipient of this "favor" has to spend much of the issue evading them in order to tend to his Daredevil duties. Various events take all the players to a large parade on 5th avenue, and Matt finds that this causes a bit of a problem: "Tickertape is jamming my radar. Have to listen for... There they are. Scrambling around on something rubbery maybe twenty feet from the ground."

In this scenario, the ticker tape seems to function similarly to chaff, a kind of radar countermeasure typically consisting of thin strips of aluminum or other metal-coated materials. We also see Daredevil having to rely on his hearing instead to find what he's looking for. Frank Miller is very much on team "radar as a separate sense" so far.

Complicating things is another idea that Miller was toying with in the radar department. The issue referenced above is actually set immediately after Daredevil had regained his ability to radar-sense. After being injured in an explosion in *Daredevil* #174, Matt spends an arduous three issues of commiserating, soul searching, and finally training to restore his radar. This is the arc that introduces Stick as the mysterious figure who trained Matt in his youth. Since Stick is a blind martial artist himself, and apparently in possession of a nondescript "proximity sense," this story may well have been what Miller was referencing in his interview with *The Comics Journal*.

Worth noting here, though, is that while Daredevil is desperate to find his old master, and finally does so a couple of issues later, he is still able to put on the costume and go out and fight! Matt does, however, express grave concerns about his future. In *Daredevil* #175, he confides in his college girlfriend turned assassin. "Elektra, you know how the radiation that blinded me in my youth also amplified my remaining senses, and I've depended on those senses ever since.

But mostly, I've depended on my built-in *radar*. It's my closest equivalent to sight. It's gone, Elektra! That bomb that hit me yesterday took it away. I hoped it would come back — but it hasn't."

In *Daredevil* #176, he tells his girlfriend Heather Glenn, "I'm *blind*, Heather! More blind than I've ever been. That bomb that hit me a few days ago — it robbed me of my *radar sense*! It's *gone...* [...] My hyper senses just aren't enough by themselves. Without my radar, I'm — I'm helpless... I've got to get it back, I've got to!"

Later in the issue, Matt finally tracks down Stick, and they begin their training in *Daredevil* #177. This is where Stick attempts an explanation of the radar sense. "Think yer something special, don't ya? Think yer the only guy in the world with built-in radar? Punk, we *all* got radar. It's a sense, just like smellin' or hearin.' Men have let it decay cuz they got eyes. You got lucky — yours got mutated. You got a free ride. But the ride's over. You got hit by a bomb that took it away. If you want it back, you gotta earn it."

Stick may be a man of many gifts, but a sound grasp of evolutionary history is not among them. The basic structure of the vertebrate eye is half a billion years old, and there has obviously never been a time before "men got eyes." However, it at least suggests that Miller imagined the sense to be compatible with normal human physiology, even while treating the radar as a separate sense that does *not* depend on hearing. However, we should note here that what Miller is suggesting *could* mean that he views both the radar and the supposed proximity sense of martial artists to be a sort of paranormal sixth sense, even though this seems to contradict its otherwise physical nature.

If you will allow me a quick detour, I would point out that Denny O'Neil, whose run picked up right after Miller's first is the one writer I can think of to have explicitly and *repeatedly* identified the radar as something not quite of this world. In *Daredevil* #195, by O'Neil and artist Klaus Janson, we read the following description of Daredevil taking in a scene:

"He stands still as stone then, and concentrates. It is something he does supremely well. For the bizarre childhood accident which deprived him of his sight also greatly increased the powers of his remaining senses. He listens, smells, feels temperature changes. Finally, he sends forth the mysterious para-normal signal he calls his *radar* and heeds the information it returns to him."

In *Daredevil* #196, the term O'Neil uses is "sixth sense," and that description reappears in *Daredevil* #199: "Sight is forever a colorless void, but the four other senses register everything. And a sixth sense — which he calls his radar — does a bit more than that."

On the functional side, Daredevil does quite a bit of, "long-range scanning" during this run, including through walls across the street, as in *Daredevil* #207, with art by William Johnson. Then again, if the radar is a form of sixth sense that doesn't depend on anything real or tangible – also known as magic – then neither distance nor physical obstacles should present a problem.

Miller ended his first continuous run with *Daredevil* #191, but returned in spectacular fashion with perhaps the best-known of all *Daredevil* arcs. *Born Again* featured art by David Mazzucchelli and ran for seven issues during the first half of 1986, starting with *Daredevil* #227.[2] If Miller's low opinion of the radar sense wasn't apparent during his first run, you get a better sense of it here. Not only does Matt appear relatively "low-powered" in general throughout these issues, but the radar sense is also barely addressed at all. *Born Again* may be the perfect example of how pushing Daredevil's senses to extremes is *not* necessarily what makes him great. On the contrary, these classic and nearly universally loved seven issues have our main character at his most vulnerable, both physically and mentally, and he spends most of them out of costume. Another thing that Miller mentioned in his 1982 interview certainly rings true here:

"Also, because he's blind, he's just barely getting by. I find Superman to be a very boring character because I never believe he's really in danger. Bullets bounce off of him. But somebody who could actually have trouble getting through rush-hour traffic interests me a great deal."

What makes Frank Miller's take on the radar sense interesting enough to deserve its own section doesn't just come down to the fact that his runs hold such a prominent place in Daredevil history, even though that's certainly one reason. After all, his time on the book brought the character a brand new set of fans and significantly raised his profile. Even in our own decade, you will find few self-professed Daredevil fans who haven't read Miller's work while the earlier issues are not viewed as required reading to the same degree.

Another thing that makes Miller's take on Daredevil's senses particularly interesting is that he brought one idea of the radar sense to his early work, and then later *changed his mind*! In *Born Again*, Miller doesn't noticeably reverse course so much as leave the radar sense out of the conversation entirely. However, another even later example of Miller's work comes to us in the form of the five-issue *Daredevil: Man Without Fear* mini-series, which featured art by John Romita Jr. and came out in 1993. Set apart from the main book, *Man Without Fear* provides these creators with the chance to reimagine Daredevil's origin, including the very nature of the radar sense. Here, there's no mention of it at all, though Matt is still bestowed with an unnamed ability to sense objects in space.

When this version of Stick finds the recently blinded young Matt, he begins to train him by asking him to hold out his hand and concentrate, explaining that, "The air's *filling* the room. One wall's *closer* than the other. *Feel* it. Now the *other* wall. *Feel* it." Next, when Stick rather brutally starts hitting Matt with his staff, Matt finally notices "[...] a *whisper* of movement... A *shape* — formed by *air*..." Months later, Matt is equipped with a bow, and told to "[f]ind the *target*. *Smell* it. Feel its *shape*. *Concentrate*." After many trials, a triumphant Matt finally declares that "I could *feel* it! I can feel *everything*—"

During the rest of the series mentions of sensation concern the remaining senses, with the occasional addition of feeling the air itself. In the third issue, one caption reads, "He gives himself a moment to let the *air* and the *smells* and the echoing *sounds* describe the place." It is not a stretch to say that *Man Without Fear* features no separate radar sense at all and that Matt's ability to sense objects rests mostly on the sensation of the pressure of the air.

We should note here that this wouldn't work. The average air pressure should be the same everywhere and sound causes only microscopic *local* deviations from the ambient pressure in the room. Sound is not wind. However, it might make sense to imagine that Matt might be able to "hear" the different walls in the room by using the build-up of sound near walls that we looked at in chapter six.

THE SELF-SUFFICIENT RADAR

In the 1985 version of the *Official Handbook of the Marvel Universe (OHOTMU)*, which came out a couple of years after the end of Miller's first run, writers Mark Gruenwald and Peter Sanderson provide us with two alternative theories of the radar sense.[3]

According to the first, "[...] Daredevil's brain has several regions which are able to sense consciously that portion of the electromagnetic spectrum that his brain constantly generates. The signals emanate from the 'sending region' of both hemispheres of his brain, travel outward, bounce off objects around him, and return to the 'receiving regions" of both hemispheres. This sense has been a 'radar sense,' since, by this theory, it functions comparably to radar."

The second "[t]heory is that this sense resembles sonar, and functions like the similar senses of bats and dolphins. According to this theory, Daredevil hears the faint echoes that sounds, even those created by his own body in otherwise near-total silence, create as they 'bounce' off the objects around him. It is possible that Daredevil's ability for detecting objects around him works by a combination of both these techniques."

What makes these types of sources interesting is that they are able to present us with entertaining factoids that typically lie outside the narrative scope of the monthly comics. However, the scientific explanations should probably be taken with a grain of salt. For instance, the snippets that cover Daredevil's sense of smell and hearing contain strange inaccuracies.[4] There's also the fact that neither of the writers of the *OHOTMU* had actually worked on the *Daredevil* comic, and so may have been working solely from their own assumptions gathered at some distance from the source material. What appears obvious is that no real "rule book" exists for Marvel writers to be working from. (At least none that specifies the physical nature of the radar sense.)

The fact that Gruenwald and Sanderson have to at least give us *something* tangible, even while leaving the question open, sets their predicament apart from the creators of the main book who can continue as before by keeping things vague. And so, the *OHOTMU* remains to my knowledge the only source to ever *explicitly* define the radar signal as being *electromagnetic* in nature. However, the idea that it is its own sense, apart from hearing and the other senses, comes to dominate the post-Miller era.

After *Born Again*, the reins as the ongoing writer passed to Ann Nocenti with *Daredevil* #236. Nocenti remained the regular writer until her final work in *Daredevil* #291, collaborating with artist John Romita Jr. for much of this run. D.G. Chichester came on the book with *Daredevil* #292 and stayed on until #332, writing additional issues under the Alan Smithee pseudonym for issues #338-342, and under his own name for *Daredevil* #380 which was the final issue of Daredevil's first volume.

Nocenti and Chichester present us with quite different approaches to *Daredevil* in terms of the general theme and tone of the book. These differences extend to how they each handled Daredevil's senses and radar, and the contrasts provide food for thought.

However, it's the similarities between their respective takes that put them both in this section.

When it comes to Nocenti's run, it is clear that the radar sense is very much meant to be its own sense, and not one that is ever described as paranormal or mystical. As with so many writers, we do see some confusion surrounding its nature. In *Daredevil* #236, Nocenti's first issue, with art by Barry Windsor-Smith, a caption reads, "Though a childhood accident robbed him of his sight – it heightened his remaining senses to an incredible degree. It also gave him a radar sense – more accurate than a bat's." The bat comparison seems more metaphorical than literal, however.

Later in the same issue, we read "Daredevil reaches out with his radar... and for the first time in his life -- something touches him back. He feels it tune into his pulse... beat with his beat... feels the beat quicken... [...]" If the first instance had us thinking of bat echolocation by another name, we are now left confused again, thinking that maybe something Daredevil is doing leaves him open to manipulation, in this case to the machinations of the villain Nuke.

A slightly different take comes our way in *Daredevil* #240 when Matt notices Karen in a crowd: "Out of the myriad infinite body shapes, one stands out. To his 'eyes,' it is the essence of grace, of beauty. The form is warm and red and glows all over, *colored by love.* Their love." This almost reminds you of the "world on fire" effect from the television show – more on that later! This concept doesn't become prominent however, and few things stand out on the radar front for much of the run, with the exception that Nocenti typically describes the radar as being highly "accurate," in contrast to many other writers, including her successor.

With Chichester on writing duties the radar sense continues to cement itself as its own thing, and one that can even be "unleashed." In *Daredevil* #308, with art by Scott McDaniel, Daredevil thinks to himself, "Radar spreads down and out, combining with other hypersenses to pick them out of the crowd." In *Daredevil* #338, ostensibly by "Alan Smithee," we read: "There's something else mixed in with the hyper-senses. A crude internal 'radar.' Long on form, short on detail. Just enough to point the blind man in the right direction. And lock him on target. [...] Air pushes ahead of the billy club, a sensory warning. The crusader tracks the sensation, twisting his body out of harm's way."

As mentioned previously, the radar sense continues to be presented as its own sense for the remainder of the first volume of the Daredevil book. For instance, in *Daredevil* #345, by J.M. DeMatteis and Ron Wagner, we read: "His radar-sense probes the gravesite — and he finds, to his relief, that Hellspawn is still there. Still very much dead." In the decidedly more light-hearted *Daredevil* #356, by Karl Kesel and Cary Nord, we catch Matt thinking, "No *light?* No *problem!* Thanks to my hyper-senses — and radar sense! From what I hear, it isn't exactly *Myst*-level graphics, but it picks out the collapsed tunnel just fine."

If we look at the more recent comics that have come out over the last fifteen years or so, the interpretation of the radar as its own separate sense, often accompanied by classic radar rings, has once again become the go-to standard for writers and artists.

In *Daredevil* #90 (1998), by Ed Brubaker and Michael Lark, Matt travels to Southern Europe, and has the following to say about one of his stops along the way:

"According to guidebooks, Sintra was one of the most beautiful spots in Portugal. I wouldn't know. The only thing that stood out to me were the long cobblestone streets winding endlessly uphill... and the walls of the 8th century Moorish castle that still stood on the mountain high above the village. It almost felt like the past was looking out at us, frozen in those ancient stones. Even my radar felt different bouncing off that much history. But it still worked. It still led me right to where I needed to be... And let me know that there was no one in my path."

During the more recent runs by Mark Waid, with art by Paolo Rivera, Marcos Martín, and (for most of the issues) Chris Samnee, the senses are explored in great detail, and it is clear that both writer and artists enjoy this particular aspect of the character. Still, the radar sense is very much its own thing. On the opening page of *Daredevil #1* (2011), a brief primer on the character tells us that "... he had developed a *sixth* sense, a *radar*-like *awareness* of where objects were." However, the sense appears to be physical in nature and, at the end of this very issue, we even see Daredevil challenged by radar chaff! Real chaff this time, not ticker tape, and he acknowledges as much, realizing that his unseen (and unsensed) attacker knows him. We find out in the next issue that the man attempting to bring Daredevil in is none other than Captain America.

A WORLD WITHOUT A RADAR SENSE

Something very interesting happens in the world of Daredevil at the turn of the millennium. The *Daredevil* title gets a bit of a fresh start with the new era that begins with the launch of the Marvel Knights imprint in 1998. A couple of years later, in 2000, Marvel launches the Ultimate Marvel imprint which features reimagined versions of many familiar comic book characters.

While Daredevil never got his own Ultimate Marvel title, he is featured in a few others, among them the four-issue mini-series *Ultimate Daredevil and Elektra* (2002), by Greg Rucka and Salvador Larroca, and its five-issue follow-up *Ultimate Elektra* (2004), by Mike Carey and Larroca. These stories follow Matt and Elektra as college students. A grown-up version of Daredevil had made his debut in *Ultimate Marvel Team-Up #7* (2001), by Brian Michael Bendis and Bill Sienkiewicz. Characteristic of both these imprints is that the radar sense is, well, *gone*.

Four shorter distinct story arcs kick off *Daredevil*, volume 2. The first eight issues, by Kevin Smith and Joe Quesada, are perhaps best remembered for the death of Karen Page at the hands of Daredevil nemesis Bullseye. There is no reference to the radar sense in either the writing or the artwork. When I look at the inside of the dust jacket of my hardcover collection of the first issues of the volume, it simply reads: "A tragic accident took his sight — but in return, enhanced Matt Murdock's remaining senses far beyond human limits."

Issues #9 through #15, which collect the story arc *Parts of a Hole,* are written by David Mack, who supplies some of the artwork along with Quesada, and introduces the deaf character Maya "Echo" Lopez. These issues offer some beautiful insight into Matt's inner life and have him interacting in interesting ways with Maya whom he gets to know in both her guises. But, no sign of the radar sense.

The same goes for *Daredevil* #16 through #19 which are written by Brian Michael Bendis, with art by David Mack. It's an unusual story that focuses on Ben Urich and his attempts to help a traumatized boy named Timmy who is the son of Silver Age villain Leap-Frog. With the framing of the story, Daredevil features less here than he usually would, but Ben Urich himself brings up Matt's childhood accident several times over the course of the story arc, but never with a mention of the radar. The trend continues with *Daredevil* #20 through #25, which cover the *Playing to the Camera arc,* written by Bob Gale, with art by Phil Winslade and Dave Ross.

Daredevil #26 kicks off the well-known "main" Bendis run which sees him paired with artist Alex Maleev, and the reimagination (or is that obliteration?) of the radar sense continues. At the start of *Daredevil* #27, the title page offers us the following: "Attorney Matt Murdock is blind, but his other four senses function with superhuman sharpness. He stalks the streets at night, a relentless avenger of justice: Daredevil - The Man Without Fear!"

In addition to the main title, Daredevil was featured in a handful of limited series during this period that were also published under the Marvel Knights imprint. We have *Daredevil: Father* (2004), by Joe Quesada as both writer and artist, as well as the unusual *Daredevil: Redemption* (2005), by David Hine and Michael Gaydos, which takes Matt Murdock on the road and mixes him up with a bit of true crime. The story of the missing radar sense repeats itself here.

.   .   .

That getting rid of the radar sense was an editorial decision seems beyond any reasonable doubt. And given its lack of any clear definition, one can see why this became a natural way to modernize the character. I am myself of the opinion that a separate radar sense isn't necessary, and can understand if it reads as a bit corny or old-fashioned.

However, this new status quo was not to last. The radar sneaks in through the back door in the "silent" issue, *Daredevil* #28, which is told solely through the artwork without captions or dialogue. Here, one panel shows Matt holding his hand in the air. The script which complements the issue has Bendis describe it as "Matt puts his hand up – like he is feeling the air. He is using his radar here." Judging by Maleev's artwork, the team seems to be inspired by a similar scene from the previously mentioned *Man Without Fear*, by Frank Miller and John Romita Jr. This might seem like a workable compromise, but I'm not convinced that this creative team quite knew what to do with this bit of insight. I greatly appreciated this fan-favorite run for the mature tone, the stunning artwork, and the high concept approach to storytelling. However, I must say that I found the writing of Daredevil's senses to be its major weak point.

For an example of how the radar sense could be described, we can look at one scene from *Daredevil* #43, in which Matt pays a visit to Luke Cage. Luke asks, "You got that radar of yours?" When Matt answers in the affirmative, Luke says, "Flip it on. Anyone in this building selling drugs? Anyone in this entire building *doing* drugs?" Matt now pauses for a while, seeming to concentrate, then answers "No."

I don't particularly mind Luke's phrasing here. However, I have to wonder how we as readers are meant to understand what is going on. Is Matt somehow sniffing through the building? Is he listening? Are we to believe that he can *simultaneously* be aware of the position of everyone within earshot? Is the radar suddenly not the crude – albeit useful – ability to sense the presence of objects relatively close by, but instead a form of transcendental knowledge for which there is little explanation? We may have a class one stimulus problem going on.

And if the idea was to replace the separate radar sense with something more grounded, this does the opposite.

By *Daredevil* #41, the radar sense had even returned to the intro blurb, which now reads, "Attorney Matt Murdock is blind, but his other senses function with superhuman sharpness and a radar sense. With amazing fighting skills he stalks the streets at night. A relentless avenger of justice. The man without fear!" Whether this change was due to audience demand or something else, the new addition looks shoehorned in. Is it even proper English?

By the time we get to the first issue of the run by Ed Brubaker and Michael Lark, *Daredevil* #82, this now reads as (emphasis mine): "Attorney Matt Murdock was blinded as a child, but his other four senses function with superhuman sharpness and *form* a radar sense. [...]" When we get to Daredevil #96, this has once again been updated, and now reads "radar *ability*" (emphasis mine). This is a bit ironic given that Brubaker and Lark had obviously returned to some version of the classic radar sense, as we noted in the previous section.

One obvious issue with the reimagined "non-radar" radar is that there is little to explain how Matt might actually be able to sense objects. As I've alluded to earlier, the concept of the radar as simply the combination of his remaining heightened senses doesn't quite work as a stand-alone explanation. In *Daredevil* #18 (1998), by Brian Michael Bendis and David Mack, Matt's friend Ben Urich shares his secret knowledge with us:

"I know that he was in a horrible accident when he was a boy and that it blinded him forever. And that the same accident that blinded him brought with it incredible sensory perception. Almost superhuman ability to taste, touch, smell and feel what he cannot see."

While I hesitate to read too much into this, it's worth noting that this list seems inadequate for this purpose, and it also leaves out hearing. I'm convinced that if the reimagined radar sense would have included a greater explicit emphasis on hearing – yes, I'm talking about echolocation – it would have made things easier for creators and readers alike. The idea of the heightened senses combined underpinning Matt's perception of objects is attractive from the perspective of the compensation narrative, but it doesn't make much intuitive sense.

If we are to identify the most recent example of a radar-sense interpretation that *explicitly* points to echolocation, we may have to as far back as *Daredevil* #87 (1964), by Gerry Conway and Gene Colan. This issue sees Matt and Natasha move in together in San Francisco, and Matt gets his first "look" at it when he steps inside and taps his cane(!) against the floor. He tells Natasha, "Let's see now. According to my trusty *radar sense* — this room's filled with *furniture* — and sheets?" Natasha explains that she rented it furnished. What we are given a sample of here is not just passive (as in *Daredevil* #47), but active echolocation!

For a modern take that comes very close to strict echolocation, we may instead look beyond the comics to the 2003 *Daredevil* movie, which famously sees Matt, played by Ben Affleck, bang his billy club against the railing on the subway platform in order to generate sound to navigate by. It has been my impression that most fans were quite happy with the Daredevil movie's take on the "radar sense," even when that was perhaps the *only* thing they liked. And yes, the "radar" in this case appears to be almost entirely based on echolocation.[5]

But even the scenes from the movie that "work" have some features that might come across as confusing. In the scene that introduces the particular visual effect that shows the viewer what Daredevil perceives, we see objects in young Matt's hospital room that are mixed in with people moving in the hallway outside, and then cars and jackhammers on the street below. They are all rendered with the same bluish special effects filter. Which is why I'm left wondering which of these things he is "seeing-hearing," as opposed to just hearing? Objects and people far away are not within echo range, by any measure, but are nevertheless portrayed in the same manner as those features of the environment he might be able to locate through real echolocation.

Another confusing aspect of this scene, and others, is that there is as much focus on the *incident* sound waves as on their reflections. By incident, I mean the sound which is making its first trip from its source to the object against which it is about to be reflected. But the detection of the silent object happens when the brain is able to compare the sound from the source with the reflected sound, which can not be done at all unless there actually *is* a reflected sound. Visualizing the wavefronts as they spread across the room *before* they have made contact with the wall or object they "illuminate," imposes an odd sort of logic.

One thing to appreciate though is that the moviemakers are able to communicate the often ephemeral nature of sound. Not counting annoying neighbors with too-sensitive car alarms that don't quit, sounds come, go, and change over time, as we have mentioned previously. This makes for an ever-changing scene that can be hard to parse, and this was communicated quite nicely throughout.

Looking instead at the *Daredevil* television show, we see an evolution of the "radar sense" – which is never actually described by that term – over the course of the three seasons. This is not surprising considering the change in showrunners and individual episode writers between seasons. By seasons two and three, it seems that Matt's sense of objects has come to depend entirely on his hearing. I mentioned this earlier in regard to season three, on the topic of how losing hearing in one ear would undercut the ability to echolocate effectively.

There is also a chilling scene from *"Dogs to a Gunfight."* the second episode of the second season that illustrates Daredevil's dependence on sound. After having been grazed by a bullet to the head in the previous episode, courtesy of Frank Castle, Matt is recovering at home. He is experiencing strange symptoms, such as his hearing becoming painfully sensitive before his ears suddenly give out entirely. At this point, an understandably panicked Matt stumbles across his living room and sits down with his back against the wall. As a viewer, you get the impression that he has lost all of his ability to sense objects as a result of this, and is quite literally both deaf and blind.

However, the first season of the show offered a different take on Matt's sense of his surroundings. An apparent attempt to visualize the experience of several of Matt's senses working together made its first and last appearance in the fifth episode, *"World on Fire,"* which was also the name given to Matt's subjective experience. Matt describes his inner world to his new friend Claire who has spent the night at his place after being viciously beaten by Russian mobsters in the previous episode:

Claire: "[...] I know that you're blind, but you... *see* so much. How?"

Matt: "I guess you have to think of it as more than just five senses. I can't see, not like everyone else, but I can feel. Things like balance, and direction. Micro-changes in air density, vibrations, blankets of temperature variations. Mix all that with what I hear, subtle smells. All of the fragments form a sort of... impressionistic painting."

Claire: But what does that look like? What do you *actually* see?"

Matt: "A world on fire."

First, what Claire is doing here is erroneously equating seeing with knowing, which is a common enough mistake considering how the word "see" is often used. Matt does at least impress on her that he doesn't actually *see*, so there is that, but his list of "senses" do not actually add up to more than five, or in any way expand on the senses all of us have.

While not usually included in the Aristotelian understanding of the five senses, balance is definitely its own sense and a very useful one. It informs the brain of the motion and acceleration of the body (the head really) relative to the ground in a field of gravity. That's it. I'm happy to hear that Matt can put this sense to good use, but it does absolutely nothing to convey information about remote objects.

The sense of "direction," meanwhile, isn't a separate sense. In fact, it's a perfect example of some of the pitfalls of the word sense itself, which we talked about in chapter two. Your "sense" of direction is your ability to plan and execute a route that takes you to where you want to go. It rests on your ability to make a mental map of your surroundings, and check that against your current whereabouts. Importantly for our purposes here, the sense of direction does not exist apart from the "true" senses, and it is certainly not something that can generate any kind of image.

Meanwhile, micro-changes in air density is pretty much a description of what *sound* is. The mention of vibrations suffers from the same problem. When a vibration travels through the air, a medium, at a frequency that we can hear, we call that a sound. We can sometimes feel a vibration through the air by our sense of touch, provided it is sufficiently intense. We can also feel the vibrations of objects by touch when we come into direct contact with those objects. At no point, however, have we created a new sense of vibration beyond the ones we already have.

Blankets of temperature variations is a tricky one. On the one hand, yes, of course we can sense those, thanks to the thermoreceptors in the skin. You feel the warmth of a nearby fire or the draft from an open window. On a sunny day, you can immediately feel the difference on your skin when you step out of the shadow and into the light, and vice versa. Matt Murdock would make use of all of these cues, and we can happily bestow him with a heightened sense of hot and cold, such that he might more easily be able to detect a *change* in temperature.

However, as mentioned in chapter three, the step from this to being able to form *images* from the temperatures of remote objects is astronomical. Given the repeated use of this kind of temperature-sensing in the first season, as in the scene with Stick and young Matt in episode seven, it does however appear that a kind of heat-vision was central to the understanding of the "radar" in the *Daredevil* show's first season.

I do have to give the season one writers credit for actually attempting to explain how all of the senses combined could create something greater than the sum of its parts. However, examining each of those parts again illustrates how they don't quite fit together. As conclusively proven a long time ago at Cornell University, it is the sense of hearing that stands far above the rest when it comes to replacing at least some properties of sight.

SOUND WAVES VERSUS RADIO WAVES

Echolocation as the natural solution to the "problem" of addressing the nature of Daredevil's radar sense has a lot going for it. Even though it doesn't exactly fit with the notion that there are literal waves of *something* flowing from Matt's head, it seems to lie closest to what Stan Lee, Bill Everett and other people at Marvel had in mind from the outset. The idea was obviously to create a hero based on the notion of the sensory compensations of blindness, including nods to "facial vision," and then dial it up a few notches with the help of traditional Silver Age radiation.

To the extent that we care whether Daredevil's power set retains these connections to something natural – albeit vastly exaggerated – then human echolocation makes the most sense. That this ability is also apparently disrupted by excessive noise and commotion also lines up nicely with the idea of Daredevil extracting spatial information from reflected sound in the environment.

The two alternative solutions we might look to would either be a bona fide sixth sense, in the vein of Denny O'Neil for example, or the presumed (but never stated) literal radar that we get from the name alone. The latter would then depend on an electromagnetic signal.

Let us briefly think about what Daredevil having a literal radar would entail. It would, first of all, require the production of a signal. There *are* animals that can generate their own electric fields, and use these to detect nearby objects and prey by the ways the field is distorted by conducting and non-conducting materials respectively. The only problem is that these animals are all fish – air is a lousy conductor of electrical currents – and that you still need specialized tissues in order to do this. Species of so-called "weakly electric fish" usually have an organ in the tail for the purposes of generating electricity, consisting of cells that are evolutionarily related to muscle tissue. The sensing of the electrical signal is more widespread along the body, as one might expect, as this allows the fish to detect objects relative to their own bodies.

Could we conceive of a version of an organ similar to that of these fish that produces oscillating charges that, in turn, generates an electromagnetic field? Could we then decide to place this organ inside Daredevil's brain, like some kind of innocuous tumor? Possibly. However, we might run into even bigger problems when it comes to receiving the reflected signal as it returns to the brain. Organic matter, including the brain and other tissues, interacts weakly with the relevant portions of the electromagnetic spectrum (presumably somewhere in the range of radio waves to microwaves), so finding a molecular mechanism that could actually *detect* this signal and pass it on to the central nervous system would be tricky, to say the least.

As we learned in chapter three, capturing any of the things we can sense requires specialized sensory receptors endowed with specialized micro-machinery in the form of receptor proteins. Shooting actual radio waves at the brain is about as helpful as popping off the top of the skull and aiming a flashlight at it. The brain doesn't sense light; the *eyes* do. The brain then *perceives* the light when the signal is communicated through the connections which link the eyes to the relevant parts of the brain.

I doubt my reservations will convince anyone who is married to the idea that Daredevil's brain creates and detects literal radio waves, and you could argue that we should file this away under "miracle exception that we don't need to explain." We absolutely could. However, I will say that it is not the elegant solution fans and creators might think it to be, and it seems to be a pretty radical – and to my mind *unnecessary* – departure from the foundational premise of heightened human senses. There is also the additional conundrum that if Daredevil's ability to navigate rests on the emissions of a physical signal, any supervillain so inclined could simply build a Daredevil detector!

·   ·   ·

But does simple echolocation really have what it takes? This is the doubt expressed by some fans when this subject is brought up. Needless to say, we probably need some of that comic book magic whenever we let *any* character loose on the streets of Marvel's version of New York City (to say nothing of outer space or any of the realms that make up the Marvel Universe). This also goes for characters that ostensibly have no powers at all!

However, when comparing a sound-based system of navigation to one based on an electromagnetic signal, the idea some people have is that radar by necessity works both faster and with higher spatial acuity. Neither one of those arguments holds up. But don't just take my word for it; let's take a look!

Real-life radar technology commonly uses radio waves, the waves that occupy the least energetic portion of the electromagnetic spectrum, *or* some portion of the microwave spectrum which lies just above radiowaves. These waves have long wavelengths, roughly between 1 m and 1 cm if you look at the technologies that are actually in common use for remote sensing. Different applications use different portions of the spectrum, and these portions are called bands.

The commonly used X-band, for instance, has a wavelength of 2.5-3.75 cm (2.5 cm is approximately one inch). As you will recall, the significance of the wavelength in this context is that it sets the limit for how large an object, or a particular feature of an object, has to be in order to be resolved. As mentioned previously, any object smaller than the wavelength of the signal cannot be detected.

The very same thing holds true for sound, so let us compare the two. A sound of 10,000 Herz, well within a young person's hearing range, has a wavelength of about 3.4 centimeters. That is comparable to the X-band microwave "radar" we already mentioned, and well below the wavelengths of most, if not all, radar applications. Higher frequencies of sound translate into even shorter wavelengths (which is how echolocating bats can detect small insects). This means that hearing is at least no worse than radar in this respect, and potentially a bit better if we assume that Daredevil can hear higher frequencies.

What about detection speed? If Daredevil is supposed to catch bullets that move faster than sound, wouldn't he need a signal that moves at the speed of light, as all electromagnetic waves do? Well, this ability didn't start out as Daredevil detecting bullets in real-time, but as him audibly detecting a heightened pulse rate, or a gun being cocked to indicate that someone was about to fire a gun.

From my reading of the comic book record, it seems that Marv Wolfman was the first writer to suggest that it is the *radar sense* that detects when a shooter is about to fire, an idea that the next writer in line, Jim Shooter, picks up on as well. However, even these writers have Daredevil "read" the shooter, not the bullet moving in real-time.

In *Daredevil* #125, we read that "DD, ol' buddy — go home for the night and thank your lucky stars — that your trusty radar sense picked up the Copperhead's pulling of the trigger — and that your hands are faster than his eyes!" In *Daredevil* #141, co-written by Wolfman and Shooter, with art by Gil Kane and Bob Brown, Daredevil thinks to himself while encountering Bullseye, "Actually I'm 'reading' his *movements* with my *radar sense*! I can figure out where he's aiming almost before he does, and detect the slight movement of his *finger* on the *trigger* —!" Of course, this is also the issue where Daredevil hilariously detects a paper plane coming at him in time to narrate its approach, but not in time to duck. Not one of his proudest moments!

But, if we *did* want Matt to be able to track bullets as they're moving, real radar sounds like the winner, right? It would, if the speed of the *stimulus* was what actually limits our ability to detect a speeding bullet. If it were, you could even argue that regular eyesight is better than radar since our eyes don't have to send out a signal and wait for it to reflect back. However, what actually limits a real person's ability to track a bullet is the speed required to react and execute an appropriate set of movements. There is simply not enough time for our nervous system to register what is happening, plan a response and send the correct signal to the relevant muscles. And with this being the limiting factor here, it's worth pointing out that we actually react faster to *sounds* than we do to events in the visual field.

One thing that's interesting though, is that when you look at skilled athletes, such as baseball or tennis players, you will often find that they are able to perform feats that might seem impossible, given what we know about the limits of the nervous system. The explanation for how this is possible comes down to their ability to read the game and actually pre-empt the actions of their opponents. This lines up well with the theory that Daredevil can deflect bullets by "reading" the shooter.

A final thing to address is the scenario where the radar sense turns into "X-ray vision," such as in the scenes where Daredevil detects people and objects on the other side of a wall. In this case, you have to wonder what kind of signal has just the right properties to first pass through a wall, then reflect off a human body and reveal its dimensions to the point where it is *recognizable* as a human body before the reflection comes back through the wall a second time where it is received by Matt.

Use one kind of signal at a particular frequency, and it never makes it to the other side of the wall. Another kind of signal might pass through the *body* just as easily as the wall (or more so), rendering the body just as "invisible." The thing to ask ourselves is whether radio waves would be better than sound in this regard, as well as whether this is an ability that makes much sense in the first place, at least to the extent occasionally seen in the comics.

Both the wavelength and the type of energy (electromagnetic waves or sound pressure waves), affect how a waveform interacts with different types of matter. If the energy is absorbed by the wall, the signal is of no use beyond that point to anyone waiting to get a signal back. If the signal is reflected, the surface will obviously be revealed, provided that we have a means of detecting it. If *all* of it is reflected, however, you obviously can't expect any of it to pass through to reflect off of any other object *beyond* that surface.

Low-frequency radio waves easily pass through both brick, cement, and people. The fact that these kinds of waves can carry information far with little obstruction is obviously a benefit to their use in communications technologies. Moving up the electromagnetic spectrum, we find long-range radar applications that are reflected off the kind of surfaces that we use these kinds of waves to detect but can penetrate fog and rain with only minor losses to absorption. Moving up further into the microwave portion, you find technologies that may not be suitable for long-range sensing because too much will be absorbed by the gasses and water vapors of the atmosphere.

From the perspective of Daredevil's ability to detect things, it should be clear at this point that there are trade-offs at each point along the spectrum in terms of what one might conceivably be able to detect, and in what detail. Being able to "see" through plastic sounds great until you consider the full implications, which include having the plastic rendered virtually invisible. This doesn't completely rule out an ability for Daredevil to sense things through walls, but I hope to at least get the point across that a number of conditions have to be met for this to work and it is virtually impossible to imagine without losses of energy, and thus information, along the way.

What about sound then? Well, anyone who has ever seen a sonogram knows that you can use sound to build up an image of a growing fetus inside the womb, though it should be stressed here that these kinds of ultrasound technologies rely on frequencies that *far* exceed anything found in nature. But why even go that far? Anyone who has ever had to listen to their neighbor's party will know that even ordinary sound travels through most walls. But – and this is quite useful – not all sound frequencies travel equally well through solid objects.

In real-life situations, we are usually surrounded by sounds of a wide mix of wavelengths, and natural sound sources hardly ever consist of just one frequency. This opens all kinds of opportunities for using the difference in the overall *spectral* quality of a reflected sound, and its original source to reveal something about the reflecting surface. This is precisely what underlies the ability of real echolocators to determine the shape, placement, and even composition of objects around them.

A nearby bush will differ from a street sign in the way it interacts with incoming sound, preferentially absorbing and reflecting different wavelengths. The two echoes will differ both in overall sound intensity - a metal sign is a better reflector - and in the mix of frequencies returning to the listener. This is the best auditory analogy we have for the colors of the visual world.

This also means that something like an empty cardboard box can be expected to reflect sound differently than an identical cardboard box filled with something dense, like sand. Of course, this all hinges on the cardboard box letting some of the sound through in the first place. If we switch out the cardboard box for a metal coffin and make the metal thick enough to reflect all incoming sound, then the trick wouldn't work.

Both sound and electromagnetic radiation that lies outside of the small portion we call visible light open up different ways of obtaining images of objects than what we are used to. In both cases, how air vibrations and incoming photons interact with matter has to do with the physical properties of the signal and surface under observation. A glass window transmits visible light while reflecting at least a portion of most sound above a certain frequency.

In terms of whether traditional "radar" beats a sound-based explanation for Daredevil's powers, we have a tie, and I would generally recommend that creators tread carefully when it comes to Daredevil actually sensing anything other than regular sound sources through a wall.

It should be clear at this point that no cohesive view of the radar can account for everything we have seen on the page of the comic. Writers appear to have gradually settled on the idea that the radar is its own sense that takes the form of literal waves flowing out of Daredevil's head. What these waves *are*, nobody seems to know. Going as deep into the "archeology" of the radar sense as I have in researching this book, I had actually expected to be able to nail down a particular point in time where the radar finally and explicitly became the literal radar that its name suggests it to be, but this has proven harder than expected.

Of course, one thing the concept of the radar as waves shooting out of Daredevil's head has going for it is that, well, there's at least a signal there. (Though we have a bigger problem trying to account for what kind of exotic organ would decipher the echo.) The opposite is true for the radar as echolocation account where we have two highly evolved organs – they're called ears – devoted to receiving and deciphering sound and passing the information on to the brain. However, there is no indication that Daredevil actively produces sounds, other than the incidental ones from moving around.

Ultimately, fans will differ on which interpretation they prefer, and, to get back to my original point, it may not matter much from a storytelling perspective. I personally find the idea that Matt possesses *human* senses, which are *heightened*, to be the most satisfying. There are plenty of sound sources, generated by the main character or external objects, that could conceivably provide the basis for navigating by sound, especially when you consider Matt's dramatically heightened hearing.

However, there have also been certain patterns to how it's been used and understood that hold up regardless of what we imagine the radar sense to be. We will look at this next, along with how the artists specifically have created the look of the radar sense on the page, which provides another window into Daredevil's inner world.

CHAPTER 11
THERE IS SOMETHING IT IS LIKE TO RADAR- SENSE

 "I really *am* blind. From a childhood accident that heightened my *remaining senses* — and added a sort of... *radar sense*! It gives me an idea of an object's *contour* — but that's *all*."

"Let me get this *straight*. You only see things in *outline*... so you decided to put on a *costume*, run around roof-tops and fight *supervillains*?"

"I am an endless contradiction that'd *never* stand up to cross-examination, Foggy. Always thought that was part of my *charm*."

Matt Murdock explains the radar sense to friend and law partner Foggy Nelson in *Daredevil* #353 (1964), by Karl Kesel and Cary Nord

If the previous two chapters were meant as a general tour of the *nature* of the radar sense over time – and across different media – this chapter will focus more on the *experiential* side of things. While some overlap exists between these two radar sense "disciplines," the question of *what it is like* to be having a radar sense experience cuts across previously identified categories.[1]

The imagined "qualia" of the radar sense is also intimately tied to how the artists of *Daredevil* have chosen to picture it on the page. So far, I have devoted a considerable amount of attention to what the writers have had to say about the radar sense, whether through caption box editorializing, or Matt's own reflections on the topic. But this becomes an exercise in "telling," not "showing," and we will spend the middle part of the chapter looking specifically at the art of the radar sense in the comics.

Line and color artists have had to devise ways of creating two-dimensional images that attempt to convey a three-dimensional experience naturally devoid of several of the features central to the very concept of an image. The solutions they come up with have also had to hit two important notes with readers. On the one hand, what is depicted on the page needs to evoke enough of the "otherness" of Daredevil's spatial world to be worth the effort while never straying so far from visual norms that this world becomes indecipherable to the reader.

Considering the futility of rendering the radar perspective under these conditions, a case could be made for not even trying. Gene Colan is a good example of someone who rarely included depictions from the radar sense point of view, so it is certainly not something that has ever been mandatory. On the other hand, Colan was also responsible for one of my favorite takes on the radar, in a scene that depicts Matt's first encounter with Natasha Romanova, and which we will return to shortly.

On the topic of *not* showing Daredevil's perspective, we should also note that many of *Daredevil's* writers have had an oddly impersonal approach to relaying Matt's experiences with the radar sense, as if there were nothing at all it is like to radar-sense. Take, for instance, the following examples from two issues written by Marv Wolfman, with art by Bob Brown and Klaus Janson.

In *Daredevil* #132, Daredevil fights Bullseye at the circus and thinks to himself, "Infallible aim is *right*... only my *radar senses* help me *dodge* his missiles. A *normal* man could *never* hope to *calculate* their trajectory in time — let alone move *fast enough* to avoid them." What is happening here is that Daredevil appears to be performing some kind of *data processing* in order to evade the objects Bullseye is peppering him with. But ask yourself this: If someone throws a ball in your direction, and you prepare to catch it, would any of it involve you doing math in your head? Probably not. Instead, you have a perception of the ball's dimensions, along with its location and movement, and *that* is what you are conscious of. Your brain is performing all kinds of processing to allow you to do this, but you're not typically conscious of what is happening under the hood, so to speak.

A few issues earlier, in *Daredevil* #128, when our hero is out looking for the Death-Stalker, he notes that "[...] since my *radar-sense* can't seem to *spot* Stalky anywhere in this park — I can *concentrate* on finding out who *this* guy is." This too is a pattern we have seen often over the years, where the radar seems to be at once a fairly long-range general-purpose sensor, and something rather impersonal. As if this were a device he might carry in his pocket or a process that runs in the background, as opposed to a true form of perception.

Neither of these scenes really point to anything that is unique to, or even typical of, this particular run, but they are good examples of a handling of the radar sense that has been quite common generally. While Daredevil may, on rare occasions, treat his other senses this way as well – mentioning that his *hearing* is detecting something – this is not how any real person talks about what they see, hear, or smell. With the radar sense, this impersonal portrayal is everywhere. *It* spots something for him, tracks something, picks up on some movement, or informs him of something pertinent. This makes the radar sense sound either like a smartphone app ("my GPS is telling me to take the next exit!") or something vaguely like the spider-sense, alerting Daredevil to whatever the other senses fail to catch.

Another common take on the radar experience that is somewhat related to this impersonal way of sensing is when it is portrayed as the *active* piecing together of a scene. Rather than performing calculations and making impersonal deductions, these scenes see him *visualizing* what is happening around him. If there exists a raw sensation of radar-sensing, it is one that is amplified or embellished by Matt's imagination, according to this account.

For instance, *Daredevil* #147 by Jim Shooter and Gil Kane, has a caption that reads, "A man who can only visualize the drama below in his mind's eye from vague radar-image silhouettes he senses, and the intricate landscape of sounds he hears…"

There are two other scenes, both written by Brian Michael Bendis, that come to my mind as particularly striking in this regard. One is from *Daredevil* #47 (1998), with art by Alex Maleev, which sees Matt meet with (his future wife) Milla in his office. The second is from a very similar sequence in *Ultimate Marvel Team-Up* #7 (2001), with art by Bill Sienkiewicz.[2] Looking at the first example, we see a sequence of four panels of Milla's face – the first almost completely drowned in red – this run's color of choice for indicating Daredevil's perspective – and the last rendered in typical non-radar fashion. The full visual transformation is accompanied by Matt's thoughts:

 "Jasmine. She's killing me with jasmine. (In a good way.) How do girls know how to smell just right? The strawberry in her hair. The jasmine on her skin. The vanilla on her feet. She's got it all on just right. I know there's more to a woman than smell. I know. But with my unique perspective, my view of the world through blind eyes and enhanced senses… smell is a big, big part of it. Even her heartbeat is elegant. She's nervous — embarrassed but her posture doesn't give her away. She's a blind woman — so posture isn't something practiced — it's something inherent. I let my radar fill in the blanks. So I can 'see' what my other senses can't give me. I feel her form. Her silky, shiny hair. Her precious, pale skin."

There is a lot going on here. Worth noting is that the last couple of items listed are things that *no* version of the radar sense should be able to provide. Neither the shine in Milla's hair, nor her skin tone, makes sense in the realm of the non-visual. In general though, it is the idea of "filling in" that interests me here.

There undeniably exists a gap between typical human vision and what any version of Daredevil's senses could possibly convey. I don't find it strange that Matt might, on occasion, try to paint himself a more visual scene of something in his head, given that he was once sighted. But the idea that imagination or deductive reasoning could come close to the "real" picture, in order to effectively bridge the gap between him and a sighted individual's impression of the same scene, doesn't hold up to scrutiny.

Some have argued that conscious experience, including vision, might be thought of as a form of controlled hallucination.[3] However, what separates the "hallucination" of real perceptions from the unwanted and often frightening version we associate with some forms of mental illness is that the former is maintained and reigned in by the continuous sensory information that calibrates it. In Matt's case, there is no sensory data available to calibrate his imagined internal visual world beyond what his other senses provide, and he has no way to check it against external realities. This clearly separates scenes playing out in the "mind's eye," from the constraints of real sensation.

The outside world is in many ways a construct. Not so much a social one as a biological one created by the brain and nervous system. We have touched on this before; the quality of an object that designates it as "red" is real to the extent that this quality exists independently of a viewer. The *redness* of red, however, is a product of the mind. Our senses have been shaped by their ability to improve our odds of survival in the world. Their utility rests on there being predictable relationships between external objects and events and how our brains perceive them. But a radar sense experience that has to be *imagined* or actively conjured into being by the perceiver, and cannot be checked against external stimuli, is not a true sensory experience.

In contrast with both the reading of the radar as an impersonal alert tool or an actively assembled inner vision, I would argue very strongly that there is *absolutely* something it is like to radar-sense, whether we can fully imagine it or not. And, if we are allowed to borrow some insights from the sensory substitution field, I would peg the radar as an ability – or experience – that allows for the external localization of perceived physical objects in space. When Daredevil senses the shape of things around him, this is not his drawing intellectual conclusions, but a result of something with the same immediacy and "realness" as other sensory experiences.

THE RADAR SENSE EXPERIENCED

The view that radar-sensing gives rise to qualia, or subjective experience, also finds wide support in the comics. The radar-as-pocket-calculator approach to storytelling has been common enough to warrant the closer examination I just gave it, but the story doesn't end there. On the contrary, several *Daredevil* writers have seemed to relish the opportunity to delve into the philosophical side of sensation and perception.

Gerry Conway is one writer I feel is owed a lot of credit for his attempts to explain how Matt actually experiences the world, and this was a point he revisited frequently during his run. This sets his work apart from most of the comics of the Silver Age which preceded it. For instance, in *Daredevil* #88 we find "As though through a heavy veil, he 'sees' an abrupt movement in the shadow-clouded radar darkness before him —" and "Radar sense: look once again through the eyes of a man who lives in darkness — 'see' as he sees — a dim, twilight world of shadows and shapes — a world he 'sees' only in his mind — a world now too confused to make sense [...]."

In *Daredevil* #96, we are treated to the following as Matt wakes up from surgery: "Radar-senses: How to describe them? For the blind super-hero called Daredevil — for the sightless lawyer, Matt Murdock — they provide a glimpse — if only a vague, ill-defined glimpse — of the world other men see — a world which now strains to reach him — and finally — does!"

You will note that Conway uses the word 'see' a lot, albeit in quotes, and this seems perfectly apt. We will return to the ways in which the radar sense is *not* sight in the final chapter, but I would still argue that to the degree that the radar emulates any of our typical sensory experiences, sight appears to be the most logical.

However, "seeing" is not the only thing that the radar has been likened to. Many fans will recall the meeting between young Matt and Elektra at Columbia University, as seen in Frank Miller's *Daredevil* #168.

When Matt describes his senses to Elektra, he tells her that "I was fifteen when I saved an old man's life by shoving him out of the way of a runaway truck. A radioactive canister from the truck struck me across the eyes, blinding me. Yes, it was [horrible]. But I later found my remaining senses incredibly heightened. I can hear the faintest whisper — even a heartbeat. I can smell a rose from a hundred feet away. I've even got a kind of 'radar,' which lets me feel objects around me. It's not like sight — it's like touching everything at once!"

The idea that to radar-sense is to touch everything at once is echoed by Mark Waid in *Daredevil* #1 (2011), in a segment with art by Marcos Martín. Chatting away (a bit too loudly?) in the subway, Matt tells Foggy, "The *radar sense* that came with the radiation is the gift that took the most getting used to. [...] It's not just some optic-substitute thing." When Foggy adds, "You've said it's more like *echolocation*," Matt replies "Like my brain is constantly pinging my surroundings 360 degrees. But there's a sort of tactile facet to it, as well. Radar sense feels like walking through the room and touching *everything at once*."

Both of these scenes remind us of real accounts of "facial vision" (now proven to be echolocation), in that these too make frequent references to touch, or a sensation of pressure. However, I must admit to not finding this interpretation to be particularly intuitive. If we are talking about approaching a wall or a single solid object, the experience is relatable. I just have a hard time wrapping my head around an externally located experience of spatially separated objects being "like touch," when the latter is by definition something that arises from sensations at or relatively near the skin. The more we are asking of the spatial information-processing capabilities of the radar-sense, the more plausible it seems that it would give rise to something sight-like.

There are other interesting accounts to look at on the topic of what life inside Matt Murdock's body is actually like. During the seventies, the idea of the radar sense firmly seems to settle into the now familiar pattern that what it gives Daredevil is a sense of the shape or outline of objects. Steve Gerber is a writer I would point to for portrayals of a kind of pseudo-visual ability that still falls short of natural vision in several key ways, which go together particularly well with the efforts of the regular artist Bob Brown. In fact, Gerber does a much better job of highlighting Matt's blindness than most Daredevil writers throughout the history of the comic and frequently comes up with tricky challenges for Matt to deal with.

One well-composed scene that involves nearly every sense – *and some deductive reasoning!* – comes from *Daredevil* #114, with art by Bob Brown, and sees Daredevil awaken near a swamp in the Everglades to discover Man-Thing fighting off the Gladiator and the Death-Stalker. "Fighting his way back to consciousness, Daredevil turns all his hyper-senses to assessing the bizarre scene — and finds it defies all logic! Three shapes, two humanoid... But only one audible heartbeat: The Gladiator's. The other man seems almost wraithlike, ghostly. There is something chilling about his very presence. But by far, the greatest enigma is this third shape, this crude, shaggy, unwieldy — beast? Whatever. And the strange "squooshing" sound it makes when it moves... as though its substance were... living slime. Even the odors are maddening — the nauseating stench of burning flesh — the foul, fetid smell of the monster's form — and a certain indefinable scent of death!"

Many writers have made a point of the crudeness of the radar, which absolutely has to be considered the majority take. While we rarely get a full sense of the limitations of the radar (more on this in the final chapter), when writers and artists are tasked with describing and defining things, they usually paint a picture of something quite rudimentary. This also goes well with what we know of the physical limitations discussed in the previous chapter. Even Denny O'Neil, whose radar was described as extra-sensory and would often reach impossibly far, made sure to underscore this aspect.

In *Daredevil* #210, which takes place during a rather bizarre arc featuring a group called the Kingore tribe, we get a look at a massive statue of the god Mow. When coming face to face with this alien object, it takes a little deductive reasoning before Matt realizes what it is he's "looking" at. Before correctly pegging it as a statue, Matt thinks to himself, "Large mass to my left. Irregular." Daredevil performs similar acts of piecing together an understanding elsewhere during this run. (Note, piecing together an *understanding* of what something *is*, is not the same as replicating the sighted *experience* of that thing in your mind's eye.)

For its impressive reach and powers of penetration through steel and concrete, this take on the radar also has some humorous limitations. In *Daredevil* #212, O'Neil has Daredevil note that "[t]he bushes are confusing my radar sense." In *Daredevil* #216, he thinks to himself, "Flower smell is almost overwhelming... making my nose useless. And all these floral pieces are confusing my radar sense —"

I give D.G. Chichester a hard time about how he introduced Matt Murdock's blissfully short-lived ability to read computer screens by touch (which we covered in chapter eight). However, as someone who really values "good senses writing," I also found much to appreciate during this run, and it is clear that Chichester enjoyed writing this aspect of the character. Paired with Scott McDaniel's art, which we'll get back to in the next section, there is an interesting amount of emphasis placed on the crudeness of the radar sense.

Except when there isn't. *Daredevil* #306 is set in the middle of a team-up arc with Spider-Man in which the two go up against the organ-stealing female villain evocatively known as the Surgeon General. When Daredevil and Spider-Man go on a fact-finding mission, the radar sense accomplishes one of its most absurd feats to date. Matt tells us, "*Hypersenses* spread out and focus in, searching out another instance of that foul *scent*. Something *crushed* beneath a worktable I see as a crude silhouette echoing back from the *radar* that speaks to my mind's eye." Walking up to the table, he finds a large insect crushed next to the table leg. "Tracing the *footprint* in the floorboard grit... tactile sensation searching the rubbery *carcass* of the latest victim in a timeless struggle. Radar picks out the insect's *contours*, the *imprint* stamped there." In this scene, we do see a mention of the crudeness Chichester often mentioned, but what the radar sense does here indicates the complete opposite. Matt is able to both perceive a squashed insect and, on closer inspection, detect the imprint of a key in the insect's body!

However, the above example was a pretty far cry from how Chichester typically wrote Daredevil's senses. In *Daredevil* #298, with art by Lee Weeks, Matt attempts a daring escape from a group of S.H.I.E.L.D. agents, noting that "[t]he rooftop reflects back at me all at once a crude map of blocky structures and room to maneuver. Hypersensitive hearing is more specific, pinpointing an ascending rhythm — leather soles scuffing their way up concrete steps." In *Daredevil* #304, with art by Ron Garney, we follow Daredevil around town for a nice slice of life story and get this intimate portrayal:

"Images of colliding, featureless figures echoing their way back to the top of the arch. Tension so thick he can almost feel it raising the hairs on the back of his arms underneath the supple red of his suit. Daredevil sees none of it — and follows it all. Senses drifting, focusing, then moving on again as radar comes back from 360 degrees at once — a wholly unique world view of the seemingly separate, mental pictures forming of how it might all come together."

The most obvious representative of writing which explicitly described the radar sense as oftentimes *better* than sight is that of Ann Nocenti. We will return to a couple of examples of this in the next (and final) chapter. However, I also want to offer an example of a genuinely compelling description of the qualia of the radar sense, during one of her final story arcs.

Here, Matt has returned to New York City after a trek that took him literally to hell and back. He soon finds himself confused, however, and even suffers a blow to the head. Over the course of *Daredevil* #284, he develops full-blown amnesia. In *Daredevil* #285, he has forgotten all about Daredevil and sets out to find the home of "Jack" Murdock, believing himself to be this Jack. He also doesn't understand what is going on with his eyes. Passing a fish vendor on the street, Matt gradually realizes what is different about him:

 "Fish lie thick as thieves. I see the fish. I *smell* them. So strong they *stink*, who could eat such stench? I see the fish. But I *don't* see the fish. I see them... As a map... a contour... I feel the fish scales... but their eyes are blank. Their stench, I can taste it... But color? What is their color? And the man... The man is as the fish... Shape... form... surface... but blank eyes... lipless... colorless... and that sound... that pounding... I can't really see... but I don't feel... I'm blind... but I sense vision... That pounding... like a heartbeat. My god... I hear his heartbeat!"

While it may be a stretch to think that Matt wouldn't be aware of his blindness, similarly to how Doctor Doom couldn't quite figure it out either, we can probably attribute some of the confusion to the amnesia alone. Either way, it is an interesting account of the radar experience from the inside, and the artwork by Lee Weeks helps bring it home with thin black lines on white creating a sort of topographic map.

VISUALIZING THE RADAR

Throughout the history of the comic, the most common solution to the radar sense dilemma in the artistic department has been the simple two-dimensional silhouette outline, sometimes with radar rings superimposed. As you will recall from the previous chapter, the very first radar perspective view that appeared in *Daredevil* #8 (1964) was a variation on this theme. This was also the style Colan returned to late in his Daredevil work, and the one Frank Miller relied on as well. Another artist that immediately comes to mind when thinking of the silhouette radar is Bob Brown who also used it often.

There are a few things that the silhouette outline has going for it. Beyond its expediency for the artists, it helps to clearly set apart what Matt "sees" from what everybody else sees. By its very nature, a silhouette is merely a line drawing filled in with a solid color that will obliterate all the details you would normally expect to find "inside" the lines. By providing a rather flat view, the silhouette perspective also focuses the attention on the foreground of the scene. Distant features of the background disappear, giving this perspective an intimate feel.

What the silhouette radar fails to capture is the three-dimensional nature of the space Daredevil inhabits, and which we can expect him to perceive as such. Other artistic takes on the radar have instead attempted to capture a sense of depth. Of the examples in this category, modern audiences are probably most familiar with the pink-on-black lines introduced by Paolo Rivera in *Daredevil* #1 (2011), then "inherited" by Chris Samnee in *Daredevil* #12 (2011) and used for the remainder of that run. The lines, which run horizontally "around" objects cleverly reveal their contours. Of course, as is the case with all of the graphic depictions of the radar sense, this too should not be taken literally. The pink lines are there for the readers' benefit, yet another visual metaphor, they are not what Matt actually "sees."

One example from this run that cleverly shows what the radar sense *fails* to convey comes from *Daredevil* #9 (2011), by Mark Waid and Paolo Rivera, in which Daredevil enters the underground domain of the Mole Man. In two consecutive panels, we see him go through a cavernous tunnel. The first presents his perspective, which merely depicts uneven rocky walls. The second is a reader's perspective, rendered in the more typical way, which reveals the walls to be lined with large monsters. This represents a rare attempt at bringing home the crucial difference between a world of detail and color, and one without these features.

The amount of background that is actually revealed by the Rivera/Samnee take on the radar tends to vary, but is usually fairly rich, albeit not infinitely so. This sets it apart from Gene Colan's attempt at the same thing. In *Daredevil* #81, we witness the first encounter between Daredevil and his future partner – in matters of both romance and crime-fighting – the Black Widow. Unconscious at the bottom of the Hudson river, Daredevil regains consciousness just as Natasha reaches out to save him. She notices him opening his eyes and assumes that he can see her. Our narrator puts the record straight:

"'*Sees*'? That's not *quite* the word we'd have *used* — to describe the *process* of Matt's uncanny *radar sense*... A sense that shows a looming, blurred figure — a figure that fades — lost with the losing of momentarily-returned consciousness..."

The radar panels that accompany this description are the first to truly attempt to convey a sense of depth. Natasha's outstretched hand is rendered in more detail, whereas her body – which is further away from Daredevil – consists of tight radial lines emanating from a bright center, the ends of the lines, varying in distance from the center, are showing us the outline of Natasha's shape.

The three-dimensional take on the radar also allows more detail to be revealed than the silhouette radar does. And occasionally there might be "too much." One interesting example of where the art seems to contradict the script comes from *Daredevil* #10 (2011), in a scene where Daredevil faces the Mole Man. At the time the issue came out, I received the following comment from one of my regular blog readers:

"[...] there was one part of the issue which threw me a little: The part early on where Mole Man asks; can't Daredevil see how ugly he is? Matt replies that no, he can't. Yet we're shown a panel of DD's radar mapping the contours of Mole Man's face, showing us that actually DD could 'see' how ugly Mole Man was."

I ended up addressing this in a lengthy post, where the artist

Paolo Rivera himself joined the conversation.[4] My conclusion – then and now – is that the probably-too-detailed radar perspective in this case represents a compromise between an attempt to underscore the difference between Daredevil's senses and our own, while also making sure the reader can recognize what it is our hero is "looking" at. When it comes to his ability to recognize faces more broadly, the consensus interpretation seems to be that he cannot. While Daredevil has "radar-sensed" faces on occasion, there are multiple examples showing him using his hands for this. In that sense, we should be trusting the writing over the art in this scenario.

This brings us to the final artistic take on the "three-dimensional" radar I wanted to make sure to cover. Scott McDaniel supplied the artwork on *Daredevil* for all but two issues from *Daredevil* #305 through #332, covering most of D.G. Chichester's run as the book's writer.[5] McDaniel's work on *Daredevil* is probably best remembered for its innovative page layouts, and for the introduction of the somewhat controversial armored costume that debuted in *Daredevil* #321, during the *Fall From Grace* storyline.

As the radar and science nut that I am, what *I* remember most fondly about McDaniel's art are the unusual radar panels, which featured faint white outlines of objects against all black, accentuated with larger white markings and dense collections of fine white lines to etch out an impression of shape. What is most striking about these panels is how otherworldly they appear. You can usually figure out what it is Matt is perceiving, but the recognition is not always instantaneous, and in many cases helped considerably by looking at the accompanying captions. This makes for an excellent example of what happens when you place less of a premium on the readers' ability to immediately understand what they are looking at, and instead embrace the truly strange.

There is, of course, a tradeoff to this way of doing things. You cannot have readers scratching their heads throughout, so these panels are used quite sparingly. I am sure they were also quite time-consuming to conceive and then draw. Still, I appreciate the attempt and found this take on the radar to be a good match for Chichester's writing which, as you will recall, often emphasized the crudeness of Matt's perceptions.

Another quite common visual take on the radar has been to not attempt to distort the image in any significant way, but merely overlay it with radar rings, a solid color, or some combination of the two. This does for Daredevil stories what squiggly lines around Peter Parker's head do for Spider-Man stories. It's a narrative trick to inform us that "Daredevil is using his radar sense here," or "this is the scene Matt is perceiving." It merely informs us that Daredevil is focusing on a particular scene, person, or object, and usually makes no attempt to emulate how he does this, or whether he is privy to all of the details that are presented to the readers.

This artistic device has been used intermittently throughout *Daredevil* history, but for some fairly recent and prominent examples, we have the work of pencillers Alex Maleev and Michael Lark and their respective collaborations with colorist Matt Hollingsworth. The latter likely accounts for the consistency between the two runs (Maleev with writer Brian Michael Bendis, and Lark with writer Ed Brubaker) that collectively cover most of the middle and end of *Daredevil* volume 2 (1998). During this era we saw entire panels washed in a berry red, with or without radar rings, to indicate Daredevil's point of view. In the case of Lark's work, in particular, these were not necessarily restricted to nearby scenes with shapes for him to sense, but often indicated more remote events that Matt is presumably *hearing*.

As with all of these different takes on the radar, there are positives and negatives associated with this one too. It is a short step away from not really showing the radar perspective at all, which could be construed as a good thing if we are to consider it a bit of a lost cause in the first place. On the other hand, it might invite readers to think that this is literally what Daredevil "sees," which I suspect is not what the artists actually had in mind.

· · ·

Perhaps the most unfortunate take on the radar, if the goal is indeed to visualize the strictly *non-visual*, is one we have seen more recently, and which I have jokingly taken to calling the "Instagram Filter Fallacy" radar.

For a prime example of what I mean, we can look to the art of Ron Garney, who collaborated with writer Charles Soule on Daredevil's fifth volume, beginning in 2015. Garney also worked on a couple of issues from *Daredevil's* first volume, including *Daredevil* #304 which is one of my favorite stand-alone issues. Since Garney is an artist whose work I generally find interesting, it is with some regret I am forced to declare this particular take on the radar a bit of a mess.

What's the problem? Well, it sure *looks* different from the non-radar panels. But this difference is not illustrative of anything meaningful, and certainly not anything non-visual. What we have are mostly ordinary panels with the colors switched around. An early scene from *All-New, All-Different Point One* (2015) #1, which leads into the new run, depicts Daredevil's first encounter with his sidekick Blindspot, a character who uses optic technology which renders him invisible. When Daredevil is able to detect Blindspot, the latter is understandable surprised. This is actually a clever use of Daredevil's powers since he will be effectively immune to an optical illusion of this kind. The problem is that the panel that shows us Matt's perspective, not only reveals Blindspot's shape but specifically depicts his face and the detailing of his costume in white, with the rest of the costume in red. Behind him are clearly defined skyscrapers in a maroon color, with all of the windows colored white. The buildings are set against a dark red night sky with stars(!) in it.

A scene in *Daredevil* #4 (2015) sees Daredevil coming face to face with a bomb on top of a stack of document folders. Here, the colors of the radar panel have been switched up to look more like a thermograph, using shades from across the color spectrum. The digital countdown timer has been removed from Matt's perspective, which I recall a lot of fans being appreciative of, as it does highlight his blindness. However, the lettering on the folders is still visible, which would suggest that Matt can actually read them.

If the point is to illustrate that Daredevil has mostly normal vision, but sees through a pair of goggles that merely gives most surfaces (but not all!) a colored tint, then this approach is perfectly fine. However, if this is *not* what the creative team was going for, and one would hope not, these artistic choices make little sense.

THE CONSPICUOUSLY ABSENT RADAR

I know I'm not the only fan to have asked myself, especially as a new reader: "Is Daredevil's radar sense always 'turned on'?" The answer is not at all obvious when reading the comics, especially when you consider the "conspicuously absent radar" phenomenon that I first mentioned in chapter nine. And since a big point of *this* chapter is to further investigate some of the patterns that emerged while we trekked through comic book history, let us look at the "conspicuously absent radar" in more detail.

When I mentioned that this particular treatment of Matt's senses has been common throughout comic book history, this was not an exaggeration. Whether writers are doing this deliberately, or simply following tradition or some unspoken rule of how things ought to be is anyone's guess. Is there perhaps a feeling deep down that relying *too* much on the radar sense is a bit of a cheat?

Whatever the case may be, the fact remains that Daredevil does not appear to be "radar-sensing" by default at all times. The practice of letting Matt first make references to what his other senses are detecting, before acknowledging the radar, is obviously easier to spot in the earlier issues. These tend to contain more internal monologue about what he is picking up and when, but you can detect it by reading between the lines, or panels, in many of the more recent comics as well.

For an illustration of how this played out a decade and a half into the book's publication, we can look at a couple of examples from Roger McKenzie's run as the writer. In *Daredevil* #153, with art by

Gene Colan, Daredevil finds himself falling toward the ground after an altercation with Cobra and Mister Hyde, and *actively* has to put the radar sense into action. "Wait! I can hear *wind currents* whipping around a *flagpole* below me — to the *left*! [...] No! *I overshot it!* Falling so *fast* — it's hard to focus *radar sense!* But I've got to grab it!"

In *Daredevil* #155, with art by Frank Robbins, Matt is stricken with a maddening headache that apparently prevents him from sensing much of anything. When Becky Blake enters in her wheelchair as the last applicant after a long day of interviews for the vacant position as Nelson & Murdock's secretary, Matt initially turns her away. Becky herself doesn't realize that Matt is blind and assumes that this is his prejudice showing. A desperate Matt thinks to himself, "M-my God! The sound of her *voice* — even her *heartbeat*... She must be *sitting down!* And I can hear rubber wheels squeaking against the floor!" Fortunately, the two sort things out and Becky lands herself a job.

You can also find examples of "CAR" in the *Daredevil* television show. Let's look at the development early in season three, where Matt wakes to discover he has lost his hearing in one ear. You might have noticed that Matt doesn't immediately seem to realize that he can no longer "see"? It is only when that ability isn't there to catch him that it becomes apparent. You could certainly argue that he is confused and disoriented, but if I woke up and couldn't see, I can guarantee that it would be shockingly obvious to me almost *immediately*.

In fact, the exact scene of Matt crashing out of bed in the first episode of season three is eerily reminiscent of a very similar scene from the Frank Miller run which takes place during the story arc where he loses his radar. In *Daredevil* #174, Matt wakes up in the hospital to his girlfriend Heather Glenn sitting at his bedside. The two have a brief conversation before Matt hears the sound of a distant radio broadcast, and rushes out of bed, only to stumble and fall to the floor. It is at this point that he realizes that his radar sense is gone. Not after waking, and not at any point during his quick chat with Heather, did he notice that anything was amiss.

We might conclude from this that resting in bed, talking to someone he knows well, is not the kind of situation where he would need to use his radar sense. This raises other questions, of course. Why would a sense be something you turn on or off? We can't turn any of our other senses off, with the exception of sight, which is done by closing our eyes. And even with sight, it would seem odd for anyone to only keep their eyes open "as needed." While our brains work hard to process what we're seeing, this is not an effort we're conscious of.

Imagining the radar sense as the sort of literal radar (or active sonar), which requires Daredevil to consciously perform some trick of the mind to send those waves flowing from his head, would offer one explanation. In this scenario, the radar really could be a sensation that is either activated at will, or by something like the rapid movements of objects toward him.

However, imagining the radar to be a sort of hearing "subtask" may also help us explain what is going on. As you might remember from chapter six, detecting echoes and detecting the location of sound sources require somewhat opposing strategies. We would expect Daredevil and his heightened senses to be excellent at both, but there may be some switching of attention needed to optimally achieve either task. If this is the case, then letting the parts of the brain that specialize in echo detection take a break in the background when not needed suddenly makes sense. This might sound odd, but attention itself is a highly limited resource and radar-sensing may need attending to, even if it doesn't necessarily require much concentration *per se*.

Another thing to consider in all of this is that it seems reasonable for the radar sense, whatever we make of the nature of it, to occupy a very different spot in the hierarchy of the senses than vision does for most of us. We have expressions such as "seeing is believing," and "I'll believe it when I see it." Seeing, and seeing *well*, is to have access to an extremely rich sensory channel that can take in a vast amount of information quickly. While the radar sense, to quote Daredevil himself, is his "closest equivalent to sight," it is not *actually* sight.

The idea that Matt's other senses take precedence over his "not-quite-sight" is another thing that helps us make sense of the CAR situation. One of my favorite scenes to illustrate this point, and a relatively recent one at that, comes from an issue by Ed Brubaker and Michael Lark. In *Daredevil #104* (1998), Matt is fighting a losing battle against his wife Milla's insanity, artificially induced by the villain Mister Fear. When he runs home to check on Milla, he finds her nurse propped against the wall at the top of the stairs, covered in blood. But what is interesting is that he doesn't just walk in and – boom! – finds her sitting there. Instead, he first pops his head into the living room next to the entryway, then pauses to remind himself to concentrate in a panel that sees the iconic radar rings around his head. Even with this added attention to the radar, this is not what helps him put the puzzle together. Starting up the stairs, he thinks "What's happened here? That smell... Blood." And *then* the whole scene comes together.

I dare say that anyone with a typical set of senses would open the front door and find the appearance of the nurse, beaten and barely alive, to be the most attention-grabbing and scene-stealing thing in the room. This is apparently not the case for Daredevil. Instead, it's the smell of blood that hits him first, while the form of the unconscious nurse makes far less of an impression on him, and initially none at all.

One interesting concept to introduce here is *salience*, which is an important topic in attention research. Salience refers to the propensity of an object to stand out against a background by virtue of the contrast of some feature that differentiates it from its surroundings. For instance, a red apple easily catches our eye in a bowl of green apples. It is rendered salient by the contrast in color.

If we apply this line of thinking to the radar sense, it would actually explain much of Daredevil's paradoxical behavior over the decades. If what he perceives through this channel generally ranks pretty low on the "salience scale," then it makes sense that the other senses are better at grabbing his attention, while "seeing" in the manner of the radar may require more of a top-down process of active attention. Large objects and surfaces close by, especially if Matt and the object are moving relative to each other, would be more salient and "attention-grabby" than distant and stationary ones. This would allow the radar to be a decent surveillance system even when his attention is directed elsewhere.

If the radar ability depends on something like a heightened senses version of echolocation, with the appropriate brain-wiring to match, then clearly the sense of hearing on which it depends is always "on." But looking at the role of attention overall and realizing that we may be dealing with a stimulus (echoes) with a presumably low degree of salience at baseline, especially in noisy environments, would go a long way towards explaining the sketchy nature of the radar sense.

Attention also comes into play in relation to a phenomenon called (visual) crowding. To borrow the definition from one review on the topic, crowding "[is] the inability to recognize objects in clutter, [and] sets a fundamental limit on conscious visual perception and object recognition throughout most of the visual field."[6] In vision, crowding typically affects the peripheral field, which is characterized by low visual acuity.[7] Of course, the peripheral field is also almost by definition the areas of a visual scene we are not actively paying attention to, unless we're doing something sneaky like *pretending* not to look at that which currently interests us. This conflation makes it hard to tease apart whether crowding is primarily a result of the low spatial acuity of the peripheral visual field, or has more to do with limits on attention *per se*. While both likely play a role, attention appears to be the more important factor.[8]

Considering that Daredevil's pseudo-visual radar sense is characterized by a lack of color contrast and, in the hands of most writers and artists, reduced spatial acuity, crowding should logically be a factor as well, albeit one that can be ameliorated to a degree by Matt actively paying attention to different areas of the scene.

. . .

Of course, this is all highly speculative on my part. What is portrayed in the *Daredevil* comic is obviously not real, and no one like Matt Murdock exists in our own world for researchers to examine. However, any or all of these factors would make for a satisfactory explanation of the CAR phenomenon. They also help bring home the point that imagining Matt's world as walking through a lit room is likely to be wide off the mark. That Matt Murdock is able to navigate through space without walking into anything, and with a general awareness of large objects and surfaces, is one of Daredevil's more "realistic" abilities. That he would have instant access to the finer details and the identities of smaller objects is much more of a stretch. In its own ways, the comic book record supports this view. (Well, some of the time. You know how that goes.)

The *Daredevil* television show, to its credit, also features scenes that basically have Matt peel back the layers of an environment. In the first episode of the second season, *"Bang,"* we see him chase down a lead that takes him to the Meatpacking District, looking for the Mexican cartel. Among the carcasses in a slaughterhouse he enters, he finds several men, one still alive, hanging from meathooks. However, he doesn't recognize that they're there until he gets fairly close, and even then it takes him a beat or two to realize what he's up against.

Given that the smell of dead human bodies is probably drowned out pretty well by the overall stench of the place, this is not surprising. A sighted person would probably react much sooner to the contrast provided by the men's clothing and spot them from a greater distance. It's a nice detail that Matt unveils the horror of the scene a bit more gradually. Similar things happen at other times as well, to the extent they can be inferred from Charlie Cox's acting. In *"The Path of the Righteous,"* the first season's eleventh episode, Matt discovers Madame Gao's blind workers and appears to gradually be piecing together what is going on. The same goes for episode nine of the second season, *"Seven Minutes in Heaven,"* when he reaches the kids in the basement of "the farm" operated by the Hand.

The writers and artists of the *Daredevil* comic have collectively painted a picture for us that depicts Matt Murdock's inner world as a rich source of sensory information. However, when put in the position to actually describe or use the "radar sense," most creative teams have given us an idea of an object-sensing ability that kicks into high gear when needed, but which is otherwise poor on detail. It makes it possible for Matt to be aware of his surroundings, and the people and objects close by. The artists have either simply indicated that the radar sense is being used, or provided an image of what the world might "look" like to Matt that suggests something that on the one hand is clearly useful, and on the other poor on detail.

How then, is it that one common view of Daredevil, among fans and creators alike, is that he can basically "see"? And that being Matt Murdock is little more than an elaborate act? For our final chapter, we need to take an honest look at the portrayal of Daredevil's missing sense, and whether the idea of full compensation has ever made much sense.

CHAPTER 12
THE MISSING SENSE

 "The secret identity can be a *relief*, Bullseye. When I'm Murdock, I don't have to use my amplified senses to pretend I'm not *blind*."

Daredevil #191 (1964), by Frank Miller and Klaus Janson

If we return briefly to the Aristotelian conception of the five senses, there is no doubt that the "radar" sense – whatever its nature or constraints – is meant to fill the role of sight for Daredevil. You will remember our hero saying as much when describing the loss of his radar sense to Elektra and Heather Glenn during Frank Miller's first run. The comparison also pops up elsewhere.

Take for instance *Daredevil* #146, by Jim Shooter and Gil Kane, which sees our hero being struck in the head with a golf ball (thanks, Bullseye!), which causes him to temporarily lose his radar sense. Foggy is perplexed when his partner trips over a chair, pointing out that the chair was in its usual location and that Matt has the office memorized. Matt then thinks to himself: "No, I *don't* — because I've never *had* to count steps and memorize like a *normal* blind man! I've always had my *radar sense* to guide me — till now."

During her run, writer Ann Nocenti frequently made the same point in even stronger terms. In *Daredevil* #242, with art by Keith Pollard, we learn that: "The red man is blind. A childhood accident with a radioactive isotope took his sight, but gave him back heightened senses and a radar that works better than sight." Similarly, in *Daredevil* #245: "A childhood accident stole his sight, but at the same time blessed him with a radar-vision that's more accurate than eyes. It also gave him heightened, razor-sharp senses." In *Daredevil* #250, Matt goes so far as to think, "I'm blind, but my radar makes me 'sighted' — so I have to pretend to be blind!"

Matt's "pretending" to be blind is also mentioned elsewhere. In *Daredevil* #119 (1964), by Tony Isabella and Bob Brown, we find Matt at the court house, reminding himself to be careful. "Hey, slow down, fella. No matter how often you've been here in the past, you're supposed to be a blind man. Don't walk quite so confidently. 'Supposed to be?' My other senses compensate for my lack of sight so much that I almost forget I really am blind. Almost. My own private 'radar' tells me the Nelson clan has proceeded me, so..."

In *Daredevil* #1 (1998), by Kevin Smith and Joe Quesada, Matt again catches himself being careless: "Sometimes I forget to play blind in the office. Gotta watch that... pun intended." In *Daredevil* #8, by the same creative team, Matt resents having to do the same at Karen's funeral. "It's vulgar – playing the role of the helpless blind man at the funeral of one of the only people who knew the whole truth about me. I feel like a liar."

There is no denying that a lot of pretense goes into the outward display of Matt Murdock's civilian life. Nor is it a stretch to suggest that his ability to sense objects in three-dimensional space replicates *some* of the features of traditional eyesight. However, there's a wide gap in logic between noting these basic facts and concluding that the world's most famous blind superhero may not be meaningfully blind at all. As we noted in the previous chapter, writers have tended to emphasize the crudeness of the radar more often than its supposed accuracy, which makes Nocenti's take the unusual one here. However, the general notion that Daredevil's remaining heightened senses – with or without a separate radar sense – compensate fully for his blindness (and then some!) is as old as the comic itself.

In *Daredevil* #168, by Frank Miller, the young Matt tells Elektra about his heightened senses, opening with "I'm blind. But I have other abilities that more than compensate." In *Daredevil* #345, by J.M. DeMatteis and Ron Wagner, readers are told that "Despite the hyper-senses that more than compensate for his blind eyes... Despite the body trained to physical perfection... he's just a man." In *Daredevil* #352, by Ben Raab and Shawn McManus, it's "Despite a pair of essentially useless eyes... A childhood accident allows Daredevil's other hyper-senses to more than *compensate* for his lack of sight."

These quotes represent just a handful of the numerous examples found throughout the *Daredevil* archives, and the idea being expressed has become central to the common understanding of the character. Going back to the earliest comics, readers have been enthusiastically informed that there is virtually no task that Matt cannot perform with greater skill or speed than a sighted man. But the question we should be asking is whether this approach to the character makes much sense. Is this core design feature actually a thinly disguised *bug*?

Most writers and artists have made sure to imbue Daredevil's world with a certain amount of "otherness," reminding readers at regular intervals of his different ways of perceiving things. But for the most part, Daredevil is written in such a way as to make the differences between Matt Murdock and his mostly sighted fans seem largely inconsequential. And, if there is a comparison to be made, we are told that it is Daredevil that ends up on top.

There is even a handful of examples from comic book history of Daredevil being able to see, in the literal sense, that are obvious mistakes. An oft-mentioned fan-favorite example of Daredevil seeing things he should not comes from *Daredevil* #208 (1964), written by Harlan Ellison and Arthur Byron Cover, with art by David Mazzucchelli. This high-paced and very entertaining issue sees Daredevil being lured by an exploding robot girl(!) into a death maze of a mansion owned by the Deathstalker's mother. When he enters one of the rooms, he looks around to notice "That oil portrait of the old woman..." To think this scene made it past two(!) writers, the artist *and* the editor with no one catching the mistake! I suppose that is telling in its own way.

THE ODDLY-SHAPED ELEPHANT IN THE ROOM

Most Daredevil creative teams have made sure to include a scene every few issues that serves to remind readers of Daredevil's blindness, and we will return to some of them at the end of the chapter. But we first need to acknowledge that such examples are far rarer than they probably *should* be, even if we were to accept the more extreme takes on Daredevil's senses as our standard.

For all his other abilities, Matt Murdock cannot see screens, pictures, colors, or read anything visually. This is something that virtually every fan and Daredevil creator seems to be in full agreement on. Where things get weird is when creators treat the issues Matt might face as sharply defined niche problems and fail to make reasonable generalizations from them.

A scene in *Daredevil* #248, by Ann Nocenti and Rick Leonardi provides one example of how this plays out. Here, Karen surprises Matt with a property she's rented to open up an addict support and legal clinic. When Matt expresses confusion about the nature of the surprise, Karen exclaims, "Can't you see...? Oh gosh, I'm sorry. This is my eternal problem with you! Your radar makes you so 'sighted' it's impossible for me to remember you're blind, 'cept when I want you to read something!"

What strikes me as deeply odd about Karen's observation is that it fails to take into account that visual "surface information" is literally everywhere. Matt's obvious problem with accessing this kind of information is not the result of some hitherto unknown form of dyslexia. The reason he cannot read the text follows from an inability to make use of *any* information that is *only* available through the visual channel. Features in the environment that are revealed exclusively in the presence of a light source and solely to observers endowed with a sophisticated optical system (a pair of eyes) are simply beyond what Matt can perceive.

Add to this a presumed, and sometimes overtly stated, inability to discern fine detail, and the notion that Matt Murdock can basically "see" falls apart, to say nothing of the nonsensical proposition that he "sees better than all of us." As does the notion that these issues would present themselves so rarely that Karen (who was living with Matt at the time) might forget about them.

Daredevil's heightened senses can *absolutely* compensate for his lost vision in some domains, and to various degrees. Echo information – or literal "radar," if you are so inclined – can convey a sense of space and make it possible to perceive the presence and general shape of solid objects. A heightened sense of hearing would also make the objects that are themselves sources of sound better sources of information. Combined with a more deliberate and efficient use of the sense of smell, it makes sense that places, objects, and people become recognizable and identifiable to a reasonable degree. But these channels of information leave out a huge chunk of what most of us take for granted, and it is quite a stretch to think that this would only present a problem in a very limited set of circumstances.

If Matt's *de facto* blindness has often been the elephant in the room, as I would argue, it is an oddly-shaped kind of elephant that can be disguised or hidden from view by the use of metaphorical filters and carefully chosen camera angles. In the case of the *Daredevil* comic, what helps keep the elephant in the background (when its presence is unwanted) are some biases or fallacies that stem in part from the way superhero stories are told, and even from certain constraints of the comic book medium itself. Needless to say, these are not on any officially recognized list of cognitive biases, but we are inventing a new discipline here...

.   .   .

The Daredevil Bias: You may be wondering how Daredevil could possibly make for a believable superhero if he were to be portrayed as blind to the extent that I would argue is appropriate. Even with his heightened senses, some would suggest that the very concept of a blind superhero is inherently ridiculous. In my view, the basic premise works, and Daredevil is surely one of the more believable superheroes out there. One important reason why Daredevil "works" is that the superheroing part of the equation is *not* where we run into the biggest issues. What Daredevil does in costume actually plays to his strengths while shielding him from many of his limitations.

Readers who have followed me here from *The Other Murdock Papers* may recall that I have often pointed out that it is Matt Murdock, in his civilian guise, who gets to deal with the kinds of situations where his blindness would present more of a challenge. "Daredevil" would absolutely have his own set of obstacles to overcome but to a lesser degree. Tracking people to dark alleys and beating them up, or pressuring them for information, is something that Daredevil is great at and which need not depend on anything close to 20/20 vision.

While research shows that partially sighted athletes have a competitive edge over totally blind athletes even in famously blind-friendly sports like judo, it is still the case that much of the close combat Daredevil engages in doesn't strictly depend on having particularly good vision.[1] Jacob Twersky, whom we've met a few times already, was captain of the wrestling team at City College, and a state champion while competing against sighted wrestlers. Ninja acrobatics is a few steps up from wrestling, to be sure, but it need not count as a visually demanding task if we allow for other ways of perceiving general shapes in three-dimensional space.

What then are some of the things that "Daredevil" *doesn't* have to do? Well, it wasn't "Daredevil" who had to sit through years of college and law school. "Daredevil" doesn't have to research legal cases, pay his bills, go to the grocery store, find that birthday gift for Foggy, or transport himself over greater distances than his billy club can take him. This all falls on Matt Murdock. In his civilian life, he has to interact with regular people he is not beating up for information, and generally exist in a society where there is a truckload of incidental visual information that he is not able to see and that his other senses really don't make up for.

The reason people rarely think of these situations is because they are generally not something you would see featured in the comic. You see more of them in the *Daredevil* television show, but even then we have to live with the fact that following Matt to Barney's so he can shop for a new business suit does not make for riveting entertainment.

Because *Daredevil* is a superhero comic, much of the spotlight naturally falls on the Daredevil persona. That is not to say that Matt doesn't appear frequently as himself, out of costume, because he does. One of the things I love about Daredevil is that so much of who Matt is, as a person, is allowed to shine through even when he's in costume as well. However, his activities as Daredevil are clearly overrepresented in the comics, whereas the activities of Matt simply going about his day are underrepresented. There is absolutely nothing inherently wrong with this, but it does mean that there is a natural bias in most Daredevil stories that serve to filter out many of the more mundane situations where his blindness might be more of an issue.

The "Yada Yada": Matt's life would realistically entail negotiating his environment in ways that are often similar to a real-life person with a significant vision impairment. But, as mentioned above, realizing this takes some reading between the lines. Or should that be panels? Not only are common daily activities rarely encountered at all in the comics, but it is also possible to "hide" things between panels, pages, or scenes.

Again, this is not an indictment of comic book creators or television writers. They cannot possibly cover everything that happens between points A and B and hope to tell a good story at the same time. The format of a monthly comic, in which a whole story – or at least a chapter – has to be told in the space of less than two dozen pages, requires that writers and artists be economical in their story-telling.

But there could still be interesting things going on that we are not privy to. If you are as old as I am, you may have noticed that I borrowed this subheading from the name of an episode of the television show *Seinfeld*, *"The Yada Yada."* The episode, which aired in 1997, featured a subplot about George Costanza dating a woman called Marcy who inserts a "yada, yada, yada" into her personal narratives at points that suggest she is hiding something. When Marcy tells George that her ex-boyfriend had visited the night before "and yada yada yada, I'm really tired today," he naturally suspects that she slept with him.

Nowhere in the Daredevil comic and beyond are scene changes quite so conspicuously edited, but I suspect I'm not the only one to have asked myself questions like, "How did he know to go there?" or "How did he learn about that piece of information?" when reading the comic. Heck, even the all-around wonderful third season of the *Daredevil* television show featured moments where I was asking myself this kind of question, knowing that for much of the season Matt wouldn't have had access to his computer or even a decent cell phone.

This is not a question I would think to ask about most other characters, but because so much of how Daredevil accomplishes certain things are left unexplained or unaddressed, the questions linger. We can only assume that contending with Matt's blindness would necessarily have to figure into the kinds of background scenarios we are rarely shown.

The Godlike Creators: Another major issue that comes into play in any piece of fiction is that we are not witnessing real events constrained by real complications. Writers and artists – and their employers – are the literal gods of these universes. They can do whatever they think serves the story, and have it play out any way they choose. In the case of Daredevil, this might mean conveniently providing him with everything he needs to know to execute the required sequence of actions.

If the writers and artists are interested in exploring Matt's senses or blindness – and at least a few certainly have been! – they are free to do so. But they can also choose not to, and make only the most minimal of efforts to stay within reasonable boundaries. All they have to do is set up the story in such a way as to minimize any potential trouble for the main character, while also not including any scenes that directly address his blindness.

While I would personally prefer that the Daredevil "gods" felt more comfortable embracing Matt's "blind side," I would also note that having the story subtly sidestepping potential complications is at least a step up from having the character conform to the demands of the story. It is the latter that typically forces writers to produce the kind of nonsensical writing I have been critiquing throughout much of this book. Simply put: It is less grating to not be informed about how Matt knew to go to a particular building than to see him reading the necessary information on a computer screen with his fingertips.

Any collection of scenes and panels that eventually makes it into the finished narrative represents a sample of the potentially infinite ways to tell roughly the same story, and how these scenes are selected and written has a big effect on how Daredevil is perceived over time, particularly when successive creative teams make similar choices. Acknowledging the elephant to the degree that I personally consider appropriate would probably be unsettling to some readers, and would definitely challenge the status quo, but I honestly believe the fan response would be predominantly positive.

Artistic Simplicity: A final factor that might influence how readers perceive the challenge that being Daredevil might present has to do with how his environment is portrayed. Artistic styles obviously vary by artist and era, but even the more realistic art styles we find in some modern comics rarely come close to capturing real life in all its detail.

One consequence of this is that, just as the plot is curated to include only certain necessary elements, so is the physical environment we see reflected in the art on the page.

Almost regardless of the scene in question, we are dealing with locations that are much less complex than the ones we encounter in real life, and less cluttered! This makes it easier to buy into the notion that all of the objects on display are readily identifiable to Daredevil's radar sense. In fact, whenever the radar perspective is presented on the page, we are usually looking at something recognizable. You might have an outline of a plant, another one of a lamp, then a table.

One interesting scene that draws attention to a challenge Daredevil might face in a truly cluttered environment comes from *Daredevil #356*, by Karl Kesel and Cary Nord. Here, Daredevil has teamed up with Misty Knight to examine the scene of a fire and look for evidence that might exonerate the otherwise villainous Mister Hyde. Matt thinks to himself:

 "Misty moves with the confidence of a professional who knows exactly what she's searching for. I stand there like a broken Disney animatronic and hope she doesn't notice. Being blind, I'm not very good at looking for things... with my eyes. I like to think my remaining heightened senses make up for that. I start sifting through a thousand sensations, tuning them out one by one. It's just that we're looking for something that shouldn't be here. The constant murmur of the river and city... The bitter tang of charred wood... Misty's voice... Nothing personal. Anything that belongs... Anything I can account for is only a distraction. A bad memory. Hyde's oppressive lingering scent. A plank stained with a young girl's blood. Sweat and death and... and..."

Accompanying this long internal monologue are, in fact, three of those radar perspective panels, white lines on black, showing a cluttered mess that really does read as being difficult to decipher. Of course, in true Daredevil fashion, his heightened senses come to the rescue when he radar-senses a metal hatch in the floor that ends up leading to a tunnel system underneath. Matt managed to solve this problem, but the kind of "visual" crowding I mentioned in the previous chapter would realistically present a constant complication to be worked around.

BLIND MOMENTS

As I hinted at earlier, there have been instances where Daredevil's blindness has been featured in the comic and beyond. The fact that such scenes are (artificially?) rare does not mean they never occur, and some creative teams have definitely enjoyed playing with this angle more than others.

At *The Other Murdock Papers*, I have jokingly taken to calling a subset of these scenes "blind moments." The blind moment describes any scene in which Daredevil actually finds himself in a bit of a bind as a consequence of his blindness. To be clear, this is obviously not the only way for creators to remind readers that Matt is blind, or to communicate how that may affect him. Any time Matt accomplishes, by non-visual means, a task for which most of the rest of us would ordinarily use vision, he could legitimately be said to be having a blind moment.

We should also recall the different tropes creators typically rely on to raise the stakes for Daredevil in battle. We know that everything from too much noise to strong smells and large crowds can make it more difficult for him to make sense of his environment. However, for something to count as a "blind moment" under my personal definition, it needs to also put the lie to that powerful Daredevil "creation myth" of full compensation.

Let's start with the humorous example seen in *Daredevil* #101 (1964), by Steve Gerber and Rich Buckler. After an amazingly trippy encounter with the reality-distorting Angar the Screamer, Daredevil is asked to describe the man by a couple of very incredulous police officers who have no idea what he's talking about: "Can you give us a description, DD? Hair color? Eyes? Distinguishing marks? We'll put out an A.P.B. on him right away." Of course, Daredevil can do no such thing, and leaves the scene muttering something about still being "a little too shaken."

Another classic example, where Matt ultimately solves the situation by being a quick thinker, comes from *Daredevil* #144 (1964), by Jim Shooter and Lee Elias. Daredevil has found a keycard he can't identify at a crime scene and gets some help from a computer technician to figure out what it does. The man explains: "It's a magnetic key, used to identify its bearer to a computerized security system in a modern building! [...] That data is coded into the key with magnetic particles which our computer can decipher — and there's the address as you can plainly see!" He points to an address being displayed by a box on the table between them. Daredevil, who is still none the wiser, asks "Uh — yes. Are you sure there *is* such an address?" He is met with the response, "Thirteen Wall St.? Of course!" and the problem is solved.

In yet another humorous encounter from the 1970's we catch Daredevil nearly damaging his reputation as a hero, or at least as a gentleman. In *Daredevil* #119 by Tony Isabella and Bob Brown, there is a scene of Daredevil riding the subway. In costume! He is comfortably seated when he suddenly picks up on some hostile vibes from the other passengers and starts piecing together an idea of the person standing up facing him. "My radar sense is picking up someone standing in front of me... Height: Five feet. Weight: Maybe 90 pounds. Wait a second! Judging from the body temperature and pulse rate... plus the scent of an overly sweet cologne... Hoo boy, Matthew. Before you stands... the proverbial little old lady!"

Matt rapidly gets to his feet and offers the lady his seat. While I don't necessarily buy the temperature-sensing for reasons we have already covered, this scene provides a nice example of something else I've pointed out previously, that is the difference between sensing the *presence* and general shape of something or someone, and *identifying* that something or someone.

Interestingly, these are all examples of scenarios where Daredevil is at risk of exposure. It is not merely the case that Matt has to put on a kind of act in his civilian life, he has to put on a potentially much *riskier* one when he is in costume. Though this is not acknowledged nearly as often, any careful consideration of what Daredevil does, and how, should inform us that this is the case. Even as early as during the infamous Mike Murdock saga, where Matt created and impersonated how own "twin," there is a hint of such an admission on Matt's part. In *Daredevil* #28 (1964) Matt thinks to himself, "I can't let [Foggy] get friendly enough with Mike to ever realize that Matt's 'brother' is also blind!"

For a recent example of just how much tension can arise from this hint of danger, we can look to the third season of the *Daredevil* television show. In the fourth episode, *"Blindsided,"* Matt bluffs his way into Ryker's prison pretending to be his very sighted law partner Foggy Nelson. While this does in fact work – and Matt gets himself in much more trouble literally fighting his way *out* of the prison – you also get the sense that this particular act could stop working at any moment.

When Matt is sent to the nurse's office after receiving a punch to the face by the former client he is milking for information, you know that anything from a very menacing-looking eye chart to the prospect of signing a release form could leave Matt in a difficult situation.[2] In the end, the whole thing is of course a set-up by Wilson Fisk, though Matt narrowly escapes a pupil check before being stabbed with a syringe. For his part, Charlie Cox puts on an impressively layered performance as a sighted actor playing a blind guy with heightened senses pretending to be sighted.

In the comics too, there is at least one example of how going undercover as sighted can be fraught. In *Daredevil* #340, by "Alan Smithee" and Alexander Jubran, Matt learns of the death of his one-time girlfriend Glorianna O'Breen when it is reported on the news.[3] But the information reaches him in a roundabout sort of way. Matt, who is operating under the pseudonym Jack Batlin, following his own faked death (only in comics and soap operas), hears about this tragedy from the artist acquaintance Theodore Stithy who is watching his TV on mute. Confusion ensues when references are made to what is showing on screen. "There on the tube! What're you, *blind*?" Stithy demands. Matt is evasive but breaks down when he is told about Glorianna. Stithy counters with, "What are you lookin' so *tore* up about? The way you don't look at things, you ain't even *gonna miss* her *pictures!*" This later statement additionally suggests that Stithy has previously noticed something being vaguely "off" about Matt/Jack.

Other situations find Matt being able to request help or information more freely, as in when dealing with people *as* Matt, or around people who know his secret. In *Daredevil* #205, by Denny O'Neil and William Johnson, Matt is looking for a clue in some photos developed from Glorianna O'Breen's camera, and tries to have a go at making sense of them himself, but has to give up. "No good. My hyper senses aren't fine enough to discern shapes in a photograph." He next asks Becky Blake, then the secretary at Nelson & Murdock, to come into his office and describe what she sees. When nothing interesting is revealed and Becky leaves, Matt slams his fist on his desk and thinks to himself "Blast! I can hear a gnat blink at a hundred yards, but I can't tell *anything* about a photograph! I'd almost forgotten that blindness is a... a *handicap*! Even for me!"

In *Daredevil* #354, by Karl Kesel and Cary Nord, Foggy finds Matt in his office, wearing the Daredevil costume. Foggy has only recently learned that Matt is Daredevil, and is naturally freaked out by his friend's apparent lack of caution: "*Matt!?!* Y-your *costume!* Your *secret!* My God — What if it wasn't *me*? What if someone was *with* me? What if someone *walked by* while the door was open? The *windows!* Can't you see someone could look in the *windows*?!"

A cavalier Matt responds: "*Slow down*, Quicksilver! First — I'm *blind*, remember? Second — I've got *heightened senses*. I knew it was *you* and you *alone* in the hallway because I could hear your distinctive *heartbeat*. And the sun's *glare* on the windows makes it impossible to see *in* — *Karen* helped me check that." Karen, of course, had learned that Matt was Daredevil about three hundred issues before Foggy, quite the opposite of what happens in the *Daredevil* television show.

With Foggy in the know as well, he too has been able to check things or pass on specific information, in the years since.

More recently, in *Daredevil* #19 (2011), by Mark Waid and Chris Samnee, Matt asks fellow superhero Hank Pym, also known as Ant-Man, to run an "at-large" search for him related to his most recent mystery, noting that "I can't work those things." This happens while Matt is at Pym's lab for a medical check-up (following Pym's micro-surgery on his brain!) so he could either be referring to Pym's computer specifically, or to the fact that this particular database might not be the most accessible.

One choice that I really appreciated from the *Daredevil* television show was that it never strained credulity by having Matt form close working relationships as Daredevil with people who don't know that he's blind. Claire finds out his secret as early as their first meeting, in the second episode of the show, and he borrows her eyes to check the incoming messages on a cell phone shortly thereafter, and more than once. In season two, when working alongside Elektra to steal the Roxxon ledger, they get to play to their respective strengths and Matt is happy to delegate the literal paperwork to Elektra. During *The Defenders* show, Matt works closely with Jessica Jones who is naturally put in charge of the seeing business.

Additionally, after Foggy learns about Matt's secret, which is the topic of much of episode ten of the first season, *"Nelson vs. Murdock,"* the two apparently settle into a routine where Matt will still occasionally ask Foggy to fill him in, or where the latter does so unprompted. For instance, in episode six of the second season, *"Regrets Only,"* Matt, Foggy, and Karen arrive at the hospital to visit their reluctant client Frank Castle, aka the Punisher. When they enter the front door, the lobby is full of people, and Matt asks "What's going on?" While this is directed at both Karen and Foggy, with the former still being kept in the dark about Matt's abilities, this is not an unreasonable thing for Matt to be asking. He can obviously tell that the place is packed, that there's a general commotion, and it wouldn't take too long to figure out that some of them might be reporters, cops, and hospital staff respectively. But, there are certain visual shortcuts available to Foggy and Karen that makes this particular information-gathering exercise much quicker for them.

In a later scene from episode eight, "*Guilty as Sin*," a similar exchange takes place where Matt asks Foggy to tell him what's going on as Frank enters the courtroom. In this case, the question is very specifically directed at Foggy, with Karen mostly out of earshot, so we have to assume it is genuine. Foggy goes on to tell him what Frank is wearing. While we didn't see it happen, it also would have made sense for Foggy to have said something about all the signs people were holding up in the gallery. (That this kind of thing wouldn't ordinarily be allowed in a courtroom is another matter.)

Whether this pattern was a conscious creative decision or not, it seems like an obvious one. I think I speak for many fans of the *Daredevil* show when I express my appreciation for the way Matt was forced to finally and gradually accept the true friendships he was offered. The emotional support given to him by his friends is by far the most important aspect of these relationships, but that these same friends are also allowed to occasionally supply Matt with "logistical support" saves us from buying into the dubious notion, still common enough in the comics, that Matt could spend any considerable amount of time around people, as Daredevil, without being found out in some way.

THE TOOLBOX

Another way the *Daredevil* show kicked it out of the park was in the realm of assistive technology. While various forms of "blind tech," from the braille watch up, have a fairly long history in the comics, scenes that feature them have been quite rare. Not so in the *Daredevil* show or the companion show *The Defenders*. There is screen-reading software, a braille display, a talking alarm clock, a braille printer, and phones that audibly relay what's on the screen.

When you watch Matt doing office work or checking his phone, it seems obvious that assistive tech would have a role to play, but it also serves as a reminder of how rarely you still see things like this in the comics. Is it because scenes featuring assistive technology serve as a reminder of the old elephant? Or is it simply that writers and artists don't feel comfortable wading into territory they may be unsure of how to portray accurately? It is understandable if creators want to default to what they know, and staying clear of this stuff entirely is easy enough to do in the comic, for some of the reasons I've addressed already.

However, one recent run that stands out for going the extra mile was that of Mark Waid, and artists Paolo Rivera, Marcos Martín, and Chris Samnee.[4] In *Daredevil* #22 (2011), by Waid and Chris Samnee, we get treated to two full pages that detail Matt sharing some of his tips and tricks as a blind person. In my review of the issue when it came out, I described the sequence as "milestone-worthy." Here's how Matt himself tells it:

 "Being blind, here are my answers to people's questions about how I handle money. Yes, it's a pain that American paper currency is of a uniform size regardless of denomination. Used to be, because of my enhanced supersenses, I could easily 'read' a bill by touch. But each advance in printing and anti-counterfeiting technology makes that more difficult. There are smartphone apps that can identify bills through photography. Slow, but helpful. To make better time, I still rely on the old folding method. Ones flat... fins folded crosswise... sawbucks lengthwise... twenties quartered. You have to trust cashiers not to shortchange you... especially on purpose. But my most helpful tip: surrender to the 21st century and use a debit card as much as possible. It's always good to have a little cash in your pocket, but the debit card is the blind man's best friend. Except when it isn't."

The scene ends with Matt failing to withdraw money from his empty bank account, and shows him listening to the spoken instructions that relay what is on the screen. While this kind of scene is exceedingly rare in the comics, we should make note of some of the tools and devices we *have* seen over the years.

The braille watch is the oldest and, by far, the most common example of technology for the blind that you'll see in the *Daredevil* comic. Matt has been seen wearing one of these since the first decade of the comic. A braille watch is more of an *adaptive* device than an *assistive* device (though the terms are used somewhat interchangeably), in that it's an example of an altered version of a product everyone uses, not a novel device created specifically to meet a need only the blind would have. The term "braille watch" is also something of a misnomer since the tactile dots that appear on the face of the watch are not standard braille. A person checks the time by opening the glass top and feeling the position of the hands.

While it might not count as technology, *per se*, Matt has also been seen using tactile graphics on occasion. One panel from *Daredevil* #314, by D.G. Chichester and Scott McDaniel, shows Matt handling a braille version of a tactile map of the New York subway system, arousing the curiosity of reporter and confidante Ben Urich. A version of the same map appeared in the second season of the television show, as Matt was preparing to go ninja hunting in the tunnels.

In the same vein, we find various kinds of braille labels. A scene that depicts Matt making a call in *Daredevil* #1 (1998), by Kevin Smith and Joe Quesada, shows a number of in-jokes hiding being the braille on the speed-dial buttons.[5] The *Daredevil* show also featured a scene, suspiciously reminiscent of an almost identical one from the 2003 *movie*, in which Matt picks out an outfit from his closet by reading the braille cards attached to the hangers. These presumably indicate the color of each item.

Other kinds of tactile indicators are also common background elements in the show, and the set designers clearly spent a lot of time on the details here. You might be thinking, wouldn't Matt just be able to smell everything in his kitchen? If he's sniffing each item individually, sure! Though, as we've learned, smells blend together and do not form easily decipherable spatial "smell-images," which would make tactile indicators, including braille labels, genuinely useful when Matt is rummaging through the cabinets.

As mentioned in chapter eight, Matt's early encounters with screen-reading technology would have you believe that he would rather read the screen by touch! Several years later, this technology makes a comeback, as seen in *Daredevil* #2 (1998), by Kevin Smith and Joe Quesada, where Matt is working in his office. However, computers were largely missing from his life for what seemed like years. When Matt is questioned by the police in *Daredevil* #44 (1998), by Brian Michael Bendis and Alex Maleev, even the officers make a point of the fact that Matt does not appear to own a personal computer. Whether he actually did at one point and got rid of it, as the officers seem to suggest, is not something we can ever know since it never appears on the pages of the comic. Matt tries to pull one over, saying "You detectives seem to have a hard time grasping the concept that I'm a *blind* man. It makes it hard to see the screen." Kudos to the officers for not taking the bait. "They don't make computers for blind people?" Indeed they do. Or software, at least.

Finally, let us tackle the tool that does make an appearance in nearly every single issue: the white cane. In *Daredevil*, the white cane becomes both a symbol of blindness and of Matt's "pretend blindness" in that it is viewed and treated as merely a prop. Before getting to the ways in which it actually makes sense for Matt to use a white cane, we should address whether it is appropriate to view it as a gatekeeper to the blindness category in the first place.

If we begin by widening that particular category to the legally blind, that is people with visual acuity worse than 20/200 or a field of vision narrower than 20 degrees, we note that only two percent of the blind or visually impaired use a guide dog for mobility, and only an additional two to eight percent use a white cane.[6] Of course, this doesn't mean that there aren't more people out there who could *benefit* from either of those options; there most likely are. The fact that people under the age of 65 make up two-thirds of cane users, despite representing only one-third of all people with vision impairment, is probably a clear indication that factors other than residual vision dictate how people get around. The elderly may be less mobile to begin with, or more content to depend on family members, whereas a younger person can be expected to place a higher value on independence. It is also in this group we would find an overrepresentation of people who have been blind since birth or childhood.

Still, the fact remains that depending on any number of factors, a person can cross well into the legal blindness category and not have her mobility severely affected by this. Because the visual modality is used for a range of different tasks, and because visual impairment may take many different forms, you may encounter a person with macular degeneration who cannot see to read due to a loss of fine central vision, but who is nonetheless able to get around safely and inconspicuously. On the other hand, you may encounter someone with retinitis pigmentosa who is still able to see well enough in the central visual field to read a standard print book but might need to use a white cane due to severe tunnel vision.

For an example of a real-life person who may challenge some preconceived notions, let's look at the American Olympian Marla Runyan, who is legally blind due to a juvenile form of macular degeneration called Stargardt's disease. Runyan, who has also competed in the Paralympics in track and field, qualified for both the 2000 and 2004 Olympic Games, finishing in 8th place overall in the 1,500 meters in the 2000 Games in Sydney. In the 1996 U.S. Olympic team trials, she also finished 10th in the heptathlon, which shows an impressive range across several track and field disciplines. For the 2002 season, Runyan was ranked number one in the United States in the 1,500 meters, the 3,000 meters, as well as the marathon, all while competing against fully sighted athletes. Runyan is clearly an outstanding athlete by any measure, but her achievements also show that such feats are not necessarily incompatible with a relatively severe vision impairment.

What about the totally blind? Clearly, mobility can be expected to be more severely affected when a person has little to no residual vision. And this is most definitely the case, but there are examples here too that seem to defy our attempts at straightforward categorization. When reading Jacob Twersky's auto-biography *The Sound of the Walls*, I was surprised to learn that Twersky refused to use a cane until he was well into his thirties. Totally blind since the age of twelve, and very nearly so since the age of six following a childhood illness, Twersky got around all over New York mostly independently relying entirely on his remaining senses. Was this a great idea? Probably not. Twersky mentions finding himself in at least a couple of perilous situations, as well as having to withstand his mother's almost constant nagging on the topic whenever he went to visit.

Most highly accomplished echolocators *do* make use of other tools as well. Daniel Kish uses a white cane, and Edward Smallwood used a guide dog. However, the fact that it is even remotely possible for a totally blind person to get around depending only on his wit and his hearing is pretty astounding. Twersky finally decided to start using a cane after teaching its use to blinded World War II veterans, applying the newly introduced two-point touch technique, and noting that it did have its uses. But I doubt anyone would argue that this made him *more* blind than he already was.

Of course, Matt Murdock cannot be expected to be placing himself in any great danger by going without a white cane. Or, to the extent that he's placing himself in danger, this has more to do with his choice of extra-curricular activities. Why this is has to do with the very *specific* ways in which his other senses can legitimately be said to compensate. *One* problem with severe, but less than total, vision loss in terms of the effects on a person's ability to get around safely has to do with the fact that the visual system creates a perception of three-dimensional space from what is essentially two not-quite-overlapping two-dimensional images presented to the eyes. With diminishing acuity, contrast, or other problems with the quality of these images, people may find themselves less able to make accurate spatial judgment calls from what they are seeing. Additionally, shadows may be difficult to tell from real objects, and all kinds of complications may arise.

What Daredevil's radar sense does, whatever we imagine it to be, is that it very neatly sidesteps such complications. Matt will never be fooled by a shadow (he can't see them), and the same stimuli and body-stimulus interactions that allow him to perceive objects and shapes in the first place also *simultaneously* allow him to sense *where* they are in three-dimensional space. This is what makes it possible to bestow Daredevil with a "sense of space" that is at once reasonably reliable, with objects and surfaces that are sensed more intensely as he approaches them, and at the same time have him operating with very little of the visual information we get from our eyes. It's a perfect "decoupling" of contrast, color, and acuity from the perception of space and the solidity of walls and objects.

Where am I going with this? Well, for Daredevil specifically, the fact that he has little to no problem getting around without colliding with anything in his way is an extremely poor predictor of how much he "sees" on a more general level. In the real world, we assume that these things go together. For Daredevil, they do not. By extension, his non-reliance on a cane for mobility purposes is a poor metric for how blind he really is by *other* metrics.

Some of these other metrics, such as access to most visual information, provide us with a good explanation for why Matt Murdock may want to carry a cane for reasons other than simply to maintain his secret identity. What we need to keep in mind is that the white cane has more than one purpose. The more obvious one is, of course, to warn its user of unseen obstacles and detect landmarks. But it also informs *other* people of the blind person's vision status.

In Matt's case, this would save him from all kinds of needless or unhelpful exchanges with other people. Let us say that Matt wants to go to a bookstore and pick up a birthday present for Foggy. And let us also all agree that this is not the kind of environment where merely sensing the shapes of objects gets you very far. If Matt shows up there with his cane, the staff at the store would hopefully notice and help him find what he's looking for. If he were to show up without the cane and try to solicit help from the staff, he might be met with "it's over in section 15" or something else that is not the least bit helpful. Then he would have to explain that he can't see very well, which may result in his being escorted to the right section, at which point he still wouldn't be able to find the book. Cue another round of questions, and maybe even suspicion on behalf of the staff member: "What the heck is this guy's *problem*?"

This is why I found it to be a very wise choice on behalf of the *Daredevil* season three writers to still have Matt using a cane around town during episodes two and three, despite the fact that he had pretty much declared "Matt Murdock" dead and buried, and despite the fact that he might have actually wanted to lay low. Sure, there's at least one example of his playing up the blind guy angle, as seen in the scene where he enters a dry-cleaning establishment, but it is also a way to make sure that he is able to elicit information that is actually useful to *him*, and not some version of "go to this address, look for the 50% SALE sign in the window."

Much of what makes the white cane a defensible part of Matt's "act" is the sheer impracticality of his trying to feign actual sighted-ness, though this line of thinking is rarely drawn to its natural conclusion in the comics.

We should finally remind ourselves that at least part of the reason people are doubting whether Matt's blindness is "real" has to do with the expectations the average person has of what constitutes blind-ness, and what a blind person may or may not be able to do. The early comics in particular would often use the phrase "helpless blind man" to describe the kind of archetype Daredevil's creators have wanted to distance him from. And, when Matt Murdock is put in the position to strategically do his blind guy "act," he often behaves in ways that seem comically exaggerated. If blindness is synonymous with helplessness, then Matt certainly cannot be blind!

But this seems like a highly dubious and antiquated way of looking at things. The more nuanced way of approaching the situation would be to realize that the extent to which Matt Murdock is functionally impacted by his blindness depends entirely on context. But general thinking on this topic could very well be muddled by the fact that characters who "straddle" the blind and sighted categories are rare in fiction. When you have to count the previously mentioned Mole-Man, and the mid-20th century cartoon character Mr. Magoo among your rare examples, you know things aren't great.

The relative scarcity of partially sighted characters in fiction invites us to think of vision impairment as an all-or-nothing condition, despite real-world evidence to the contrary. When you combine this state of affairs with the fact that Matt Murdock is able to do things that we do not, for legitimate reasons, expect of the totally blind (or really anyone else), it is easy to see the temptation for Daredevil's creators to simply treat him as sighted for most purposes.

A big part of the reason Matt doesn't come across as "more blind" than he usually does has nothing to do with his senses being so amazingly powerful, and everything to do with a sort of narrative blindspot that has been with us since Daredevil's early days.

Some creators have indeed done a commendable job with these aspects of the character, and most writers have at least seen fit to make an occasional nod to Daredevil's missing sense. However, Daredevil's very concept, in many ways brilliant in its simplicity, carries within it this impossible goal that Daredevil be both blind, and yet somehow *not*. And this doesn't work, unless we remain content to simply avoid addressing it.

Sometimes, his senses "compensate," or take him even further, into realms of experience hidden from the rest of us. At other times, these same senses do not compensate at all. Matt Murdock is no more a "lie" than Daredevil is. Regardless of what guise he is in, Matt can be described as simultaneously having both a disability and "super abilities." Absolutely nothing forces us to pick a side, and Matt himself certainly doesn't get to choose.

Having to pretend to be a totally blind man without heightened senses is not Matt's natural state of being, but neither is having to pretend to see things he cannot or conduct himself in ways identical to someone who can see. Perhaps Frank Miller put it best when he had Matt tell Bullseye, in *Daredevil* #191:

"The hidden identity can be a relief, Bullseye. When I'm Murdock, I don't have to use my amplified senses to pretend I'm not blind."

CONCLUSION: BEING MATT MURDOCK

A few years ago, I went with a friend to an immersive experience called "Invisible Exhibition" (Osynlig Utställning). Filling an entire mid-sized warehouse on the outskirts of Stockholm, the exhibition consisted of replicas of real indoor and outdoor environments built to scale, all plunged into complete darkness. Among them were a typical apartment, a bus stop, a city square, and a few others. Our group of four or five people was safely guided through this (to us) foreign environment by a blind young woman who seemed quite amused by the whole thing.

The aim of this and other similar exhibitions is to allow sighted people some insight (no pun intended) into what it's like to be blind. Of course, a sighted person should be careful to not rely too much on the impression one might walk away with after a mere hour in the dark, without having acquired any of the skills of people who are blind in real life. Still, I found the experience valuable and interesting, and I made sure to carry my inner Matt Murdock with me along the way.

Because if we are going to build Daredevil's world, in our own minds or for the comics or live action, this needs to be where it all begins. I'm not suggesting that we *end* here, merely that starting at the opposite end of visual experience is guaranteed to lead us astray. Taking the sighted person's view of the world and merely running it through a kind of filter will never allow us to escape the dependence on light we associate with typical eyesight. It will simply be vision by another name. Instead, we have to look at other ways to support an experience of a three-dimensional external world, and other stimuli that can provide rich sources of information.

What would happen, for instance, if we turned up the volume? Especially at the low and high ends of the human hearing range. Which of the many objects obscured by the dark would make themselves known to us? There's a radio turned on, and we hear the low rumble of traffic nearby, suddenly louder and more present than it should be. But what about the sound of the turned-off TV that's sure to be in there somewhere? And the alarm system? Or perhaps the sound of the wind rustling the leaves of a nearby tree that is just a tad too quiet for the ears of the average person.

And what can we learn about the people around us? There might be the sounds of beating hearts, if we listen *very* carefully, but more obvious would be the air moving through windpipes, the slow churning of intestines, the rumble of hungry stomachs, and the smacking of a person swallowing. There is nowhere to hide in our presence.

Suppose also that our brains and ears can make useful note of the faint echoes that return to us as we move through the environment, as well as the change in pitch of the background noise near large structures. Or the "shadow" cast by the objects between us and more distant sources of sound. These avenues are open to anyone with the prerequisite skills, honed by many hours of practice, but more readily available to us, with our heightened sense of hearing and optimal neural wiring to match. The difference is in the details, quite literally. Sensitive as we are to the high frequencies that do not travel very far, but give us nice, clear echoes that reveal even smaller features. Not small enough to rival visible light as it is detected by the central visual field, but not too shabby.

Attuned as we are to all the spatial features of the soundscape, objects will begin to appear to us as occupying three-dimensional space at their respective distances from our bodies. And the experience is not one of hearing but of something else. There is the vision-like perception of shape and external localization, and even material texture. What is lacking is color, shadow, and visual contrast, which creates a different experience altogether. Perhaps what we're perceiving is reminiscent of looking at one of those 3D stereograms that trick the brain into seeing a pop-out relief of some object against an identically-patterned background when you look at the image just right. Switch the abstract patterns that create the effect for something colorless, and perhaps it is not too far off the mark. Of course, it would have to be more ephemeral, operating not under the typically stable light sources around us, but revealed by a messier mix of sounds.

At some distance away, faint echoes will no longer be available to us, but louder sounds will. And perhaps other stimuli reach us from this void as well. A light wind might be blowing our way, carrying with it a particularly interesting scent. We decide to follow it, noticing where it gets stronger. Along the way, the faint impressions of surfaces keep us safe and help us make sense of the source of the scent. Yup, it's a garbage can. Or is it? This is our very own thought experiment and since we are uniquely skilled and sensitive smellers with an unusual (for a human) propensity to notice scents it may be any number of things we've stumbled across. A dead body? The abandoned shrine of a ninja ritual? Yes, what we are smelling is clearly some brand of incense you can't buy in stores. At least not in New York (we have long since left that warehouse in Stockholm.)

Before us are objects large enough to be noticeable, but too small and indistinct to make much sense. No problem, we reach out to touch them and find small ornate vessels that are surely otherworldly in origin. Have to be. Hm, we probably shouldn't have done that. Oops. We are saved by sounds nearby. Sliding glass doors, slow-moving cars, and even slower-moving people. From the bags they are carrying come scents of apples and produce, meats, and freshly ground coffee. A grocery store.

As we focus our attention on the location, we come to perceive its solid structure. It could be the perfect spot to seek refuge from vengeful ninjas. Or maybe we'd be responsible for a blood bath?! But this is New York, surely there are superheroes around. This is clearly the AP version of the thought experiment. Let's get out of here.

I mentioned that I took Matt Murdock with me to that warehouse in Stockholm, but I'm going to let you in on a little secret: I run these kinds of thought experiments often. I have for many years now. I'll be waiting for the train, grocery shopping, or trying to find the room for the next session at a conference, and I keep thinking: What would Matt Murdock make of this? What would come more easily to him? And, what wouldn't he be able to detect at all?

If your own answer to any such question is that he'd simply take care of it by some nondescript sensory voodoo, I suggest you give it a little more thought, a few minutes here and there. Think about what you are sensing and how. What's the stimulus? What's the sense? How would you operate without one or the other? What would you do to get around such complications? What might you hear, smell, and feel if you had heightened senses? Not magical fix-it-all senses, just heightened and able to detect more of what is out there, just beyond our own limits.

These kinds of questions remind us of what it is to be human, and that there are *different* ways to be human. We share a common world, but we move through it differently, perceive it differently, and cut slightly different slices from the matter and energy that is out there to support our conscious experience of life on this planet.

This is also why Daredevil is the best thought experiment in comics. You just have to be willing to run it. And lean into it fully.

ACKNOWLEDGMENTS

I first want to thank my family for their ardent support of all of my varied pursuits over the years, including this book. They've offered love and encouragement despite knowing nothing about Daredevil (and while quite possibly doubting my sanity). On that topic, I do hold out some hope that my beloved nephews will grow up to be Daredevil fans. *Edward och Alfred, jag älskar er.*

As for the people who actually *know* who Daredevil is (and why he's great), a big thank you goes to everyone I've encountered in the fandom. I owe a special debt of gratitude to Kuljit Mithra, the webmaster of *ManWithoutFear.com*, without whom I never would have become a Daredevil fan in the first place. I will also always treasure those early days on Kuljit's message board, where I got to meet Alice, Gloria, and Francesco.

After setting up shop at *The Other Murdock Papers* in late 2007, I encountered even more people who shared my love of Daredevil and am very grateful to everyone who came to comment, engage, and ask questions. The late Aaron Kimel was among them, and he quickly became my "science muse." We had similar views on the character and how he might fit into something closer to the real world, and we would engage in long discussions on this topic. Aaron has often been in my thoughts during this project as well, and I hope I've been able to put together something he would have enjoyed.

Since the *Daredevil* television show was canceled on Netflix in late November of 2018, I have also had the pleasure of getting to know the wider #FandomWithoutFear, and I will be forever grateful for the friendships I have made as a member of the #SaveDaredevil team.

The support from the team has meant everything to me, even though finishing this book took so long that we managed to become "#SavedDaredevil" before all was said and done. I will not name you all individually – you know who you are – but would like to direct a special thank you to Shelby and Phyllis. Shelby enthusiastically read the first draft of the manuscript, alongside my professional editor (thank you, Jeff!). Meanwhile, Phyllis designed the book cover for which fellow Daredevil fan and gifted artist Monique provided the artwork.

I would also like to acknowledge "Van" Diaz and her support of all creative ventures across this and other fandoms. Van was exceptionally knowledgeable on several topics – Daredevil very much among them – and inspired me and many others to have the courage to create and share our passions with the world. It was Van who provided the quote I used for my introduction, about the status of the comics as officially sanctioned fan fiction, and I so wish she would have been around to see it in print, and to share in the many recent successes of the campaign team of which she was a core member.

Last, but certainly not least, I would like to express my gratitude to everyone who has ever worked on *Daredevil*. I'm talking about all the writers, artists (pencilers, inkers, and colorists), letterers, and editors who have overseen Matt Murdock's many adventures. The same goes for everyone associated with Daredevil's live-action outings. While I have not necessarily loved every run or narrative decision *equally*, almost everything committed to the page or screen has provided food for thought and helped create a collective idea of who this character is (and material for this book!).

I have also had the great privilege of meeting or otherwise connecting with several of the fine people who belong to this group over my many years as a fan, and it has been an amazing thrill and honor. Thank you!

NOTES & REFERENCES

INTRODUCTION: SCIENCE IS YOUR FRIEND

1. The story of how the first issue of *Daredevil* even came to be is itself a complicated one. It took many months to make it to the printer, and Steve Ditko had to step in to finish the artwork. Jack Kirby provided most of the art for the cover, and also made contributions to the character design. For a closer at this backstory, see

    Brevoort, T. (2020, December 5). *Lee & Kirby & Ditko & Everett & Brodsky: The Long Road to DAREDEVIL #1*. The Tom Brevoort Experience. Retrieved June 3, 2022, from https://tombrevoort.com/2020/12/05/lee-kirby-ditko-everett-brodsky-the-long-road-to-daredevil-1/

2. To be precise, Spider-Man's first appearance – which included the famous incident with the spider bite – actually took place in *Amazing Fantasy* #15.

1. THE LITERARY ADVENTURES OF THE SUPER-BLIND

1. Bolt, D. (2006). Beneficial Blindness: Literary Representation and the So-Called Positive Stereotyping of People with Impaired Vision. *Journal of Disability Studies*, *12*, 80-100.

2. It should be noted that the Marvel Universe is also home to other, non-super-powered blind characters, such as Alicia Masters, Ben Grimm's long-standing love interest in the Fantastic Four, and characters from the Daredevil comic itself. The latter include the blind Vietnam War veteran Willie Lincoln who was introduced in Daredevil #47 and made a half-dozen subsequent appearances, Matt's wife Milla Donovan who first appeared in Daredevil #41 (1998), and Austin Cao, a young blind man whom Matt Murdock takes on as a client in Daredevil #4 (2011). In the Daredevil comic, we are thus able to see the difference between what the creators expect of an "ordinary" blind person, and what Daredevil is able to do, in the way Matt performs around people who don't know that he is secretly Daredevil.

3. A few years ago, when doing research for a blog post, I came across a message board where a blind woman, an office worker, shared the story of how she had approached her boss about doing something to address a disturbing noise near her desk. Her boss had leaned in and asked, in a hushed tone, whether this was one of those sounds only she, a blind person, could hear. Needless to say, the woman was baffled. This example clearly illustrates how fictional tropes can affect how blind people are viewed in real life.

4. Though if we are to trust a particular line from *Daredevil* #7 (1964), by Stan Lee and Wally Wood, Matt Murdock was an early practitioner of yoga.

5. See *Marvel Comics: The Untold Story,* by Sean Howe for a reference to Miller's own account. Miller's first issue, Daredevil #158, was released on January 30, 1979 (with the cover date May 1979). An earlier printing of this book erroneously gave Miller's age as nineteen at this time.

6. The so-called Master of Kung Fu had first appeared in *Special Marvel Edition* #15, by Steve Englehart and Jim Starlin, in 1973, and had direct ties to the popular television show.

7. Although first published the same year, the Black Bat has no connection to DC's Batman. Nor did a connection exist between the 1939 version of the Black Bat and an earlier pulp hero by the same name which first appeared in 1933 in a publication by a different publisher.

8. Hecht, S., & Pirenne, M. H. (1940). THE SENSIBILITY OF THE NOCTURNAL LONG-EARED OWL IN THE SPECTRUM. In Journal of General Physiology (Vol. 23, Issue 6, pp. 709–717). Rockefeller University Press. https://doi.org/10.1085/jgp.23.6.709

9. See: Cronin, B. (2008, October 30). *Comic Book Legends Revealed #179.* Comic Book Resources (CBR). Retrieved June 5, 2022, from https://www.cbr.com/comic-book-legends-revealed-179/

    The similarities to Batman are also underscored by the clever tools the Shroud uses. His cape is lined with asbestos, which makes him fireproof against the attacks of the Human Torch in issue #5. In issue #7, we are introduced to the "bomb-a-rang," a gadget that starts out looking like a Swiss army knife crossed with a pack of gum, but folds out into an odd-looking boomerang. Together with the "parallo-mist," a titanium net, and finally a magnesium bomb, the Shroud manages to incapacitate none other than Doctor Doom.

10. Issue by Bill Mantlo, Bob Hall, and Don Perlin.

11. The 2004 edition of *The Official Handbook of The Marvel Universe* says of the radar sense: "Its resolution is not very fine, probably on the order of several feet at a distance of one hundred feet. By repositioning his head and adding input from his other senses, Daredevil is able to resolve the image of an average flag-pole (three inch cylinder) at a distance of over 80 feet."

12. Blind Faith first appeared in Justice League Task Force #10 (1993), by Michael Jan Friedman and Sal Velluto.

13. We are still a few months ahead of the release of *Daredevil* #1, as *Fantastic Four* #22 hit the stands in October of 1963.

14. Varma, Rohit et al. "Visual Impairment and Blindness in Adults in the United States: Demographic and Geographic Variations From 2015 to 2050." JAMA ophthalmology vol. 134,7 (2016): 802-9. doi:10.1001/jamaophthalmol.2016.1284

15. Finding the exact statistics here is tricky, and will differ depending on sources, definitions, and methodologies. In *Sight Unseen,* Georgina Kleege mentions ten percent as the rough proportion of the legally blind who lack visual experience entirely.

2. SENSE, NONSENSE, AND THE STIMULUS

1. Source: https://www.lexico.com/en/definition/sense (Lexico is a free Oxford Dictionary-powered website)

2. The characters on the page providing us with running commentary of their plans and actions is a common finding in comic books of this era.

3. The "savory" category, also known as umami, from the Japanese term for (roughly) "pleasant savory taste," is a relative newcomer on the list. It has also been argued that we have a special taste category for starches, see for instance:

   Sclafani, A. (2004). The sixth taste? In Appetite (Vol. 43, Issue 1, pp. 1–3). Elsevier BV. https://doi.org/10.1016/j.appet.2004.03.007

4. See Morrot, G., Brochet, F., & Dubourdieu, D. (2001). The Color of Odors. In Brain and Language (Vol. 79, Issue 2, pp. 309–320). Elsevier BV. https://doi.org/10.1006/brln.2001.2493

5. To make things more complicated, electromagnetic waves actually represent a union of an electric field and a magnetic field and the two portions actually travel perpendicularly relative to one another. If we imagine that the electric field is moving up and down along our imagined x-axis, the magnetic field will be moving "in and out."

6. The paragraphs dealing with light, sound, and odorous molecules and their respective properties have been inspired in part by the handling of this topic by Casey O'Callaghan:

   O'Callaghan, C. (2020). Senses as Capacities. In Multisensory Research (Vol. 34, Issue 3, pp. 233–259). Brill. https://doi.org/10.1163/22134808-bja10024.

7. As a reminder, photons of light can even traverse a vacuum and so need no medium at all. Sound needs a medium such as air or water (or solids) to propagate but moves as vibrations that travel at a greater speed than the medium itself (sound is not *wind*). Meanwhile, odors consist of actual matter and are not an energy waveform of any kind.

8. This scene is featured in the seventh episode of the first season, *"Stick."*

3. BODY, MEET WORLD

1. There is also the special case of nociceptors, or pain receptors. Nociceptors are generally free nerve endings that are sensitive to mechanical, thermal and/or chemical stimuli. While the sensation of pain is most definitely distinct, there is no particular stimulus that is unique to pain. Rather, pain is typically the result of "too much" of a particular stimulus that our nervous system needs to warn us about.

2. GPCRs are folded in such a way that the protein snakes its way through the membrane seven times so that three of the "loops" are on the outside and three of them are on the inside. For this reason, they are also sometimes called 7 transmembrane receptors.

3. The central role of Vitamin A in photoreception is why vitamin A deficiency causes blindness, and it remains the most common form of preventable blindness in the world. Vitamin A is also critical to the health of other structures in the eye, such as the cornea. According to the American Academy of Ophthalmology, 250,000-5000,000 children worldwide go blind from Vitamin A deficiency every year.

4. TRPV1 is a member of the TRP superfamily of ion channels. TRP stands for "transient receptor potential" and the name stems from the context in which they were discovered, in the fruit fly.

5. It has been estimated that among mammals, the size of the eye alone explains about 35 percent of the variance in visual acuity. See Veilleux, C. C., & Kirk, E. C. (2014). Visual Acuity in Mammals: Effects of Eye Size and Ecology. In Brain, Behavior and Evolution (Vol. 83, Issue 1, pp. 43–53). S. Karger AG. https://doi.org/10.1159/000357830

6. I would hereby like to dedicate this section to my own cats Murdock and Elektra who did nothing else to support me while writing this book. They did keep demanding to be fed and petted at regular intervals, however.

7. It should go without saying that getting reliable measurements of the hearing ranges of animals is fraught. You are likely to get slightly different numbers in different experimental settings, and depending on the individual animal. The numbers I have used here come from the following source:

   West, C. D. (1985). The relationship of the spiral turns of the cochlea and the length of the basilar membrane to the range of audible frequencies in ground dwelling mammals. The Journal of the Acoustical Society of America, 77(3), 1091–1101. https://doi.org/10.1121/1.392227

8. Manoussaki, D., Chadwick, R. S., Ketten, D. R., Arruda, J., Dimitriadis, E. K., & O'Malley, J. T. (2008). The influence of cochlear shape on low-frequency hearing. Proceedings of the National Academy of Sciences, 105(16), 6162–6166. https://doi.org/10.1073/pnas.0710037105

4. SENSES, MEET BRAIN

1. Edgar Douglas Adrian was awarded the Nobel prize in 1932 for his 1928 discovery that the electrical signals in the nervous system always have a certain size. "More intensive stimuli do not result in stronger signals, but rather signals that are sent more often and through more nerve fibers," to quote the Nobel Prize information page. The prize was shared with Charles Sherrington who had done work in the 1890s on muscle contractions and reflexes, and their connection the brain and spinal cord. The basic idea of the action potential is, however, considerably older and goes back to the mid-19th century.

2. In a real nervous system, including ours, there is always some spontaneous background activity going on, noise if you will, kind of like drops from a leaky faucet. But that is perfectly fine, and not a design flaw. In fact, it sets the stage for neural inhibition by which a neuron can receive instructions from a neigh-

boring neuron to decrease its base rate of firing. It is certainly useful to be able to dial the basic setting both up and down.

3. The hard problem was formulated by philosopher David Chalmers in 1995. There is of course fundamental disagreement among neuroscientists and philosophers of consciousness regarding whether the problem is in fact "hard." See: Chalmers, David (1995). Facing up to the problem of consciousness. Journal of Consciousness Studies 2 (3):200-19.

4. James, W. *Principles of Psychology* (Dover, New York, 1890).

5. Sur, M., Garraghty, P. E., & Roe, A. W. (1988). Experimentally Induced Visual Projections into Auditory Thalamus and Cortex. In Science (Vol. 242, Issue 4884, pp. 1437–1441). American Association for the Advancement of Science (AAAS). https://doi.org/10.1126/science.2462279

Roe, A. W., Pallas, S. L., Hahm, J.-O., & Sur, M. (1990). A Map of Visual Space Induced in Primary Auditory Cortex. Science, 250(4982), 818–820. https://doi.org/10.1126/science.2237432

von Melchner, L., Pallas, S. L., & Sur, M. (2000). Visual behaviour mediated by retinal projections directed to the auditory pathway. Nature, 404(6780), 871–876. https://doi.org/10.1038/35009102

6. Speaking of the sleep-wake cycle, it is quite common for people who are totally blind to experience sleep disturbances. When it comes to Matt Murdock, it has long been a pet theory of mine that an unusual sleep pattern could inspire some of his late-night activities. Fanfic writers, you may consider this a prompt!

7. A "nucleus" in neuroanatomy is simply the name of a distinct cluster of neurons. The neural structure of a nucleus is reminiscent of what you would find in the cerebral cortex.

8. Saenz, M., & Langers, D. R. M. (2014). Tonotopic mapping of human auditory cortex. Hearing Research, 307, 42–52. https://doi.org/10.1016/j.heares.2013.07.016

9. See also: Wright, H., & Foerder, P. (2020). The Missing Female Homunculus. Leonardo, 1–8. https://doi.org/10.1162/leon_a_02012

10. One indication of how much we still have to learn about the brain is that we scientists cannot even agree on how many neurons there actually are. The commonly cited 86 billion has been proposed by Suzana Herculano-Houzel, but you are bound to find other figures cited elsewhere.

11. Norman, L. J., & Thaler, L. (2019). Retinotopic-like maps of spatial sound in primary 'visual' cortex of blind human echolocators. Proceedings. Biological sciences, 286(1912), 20191910. https://doi.org/10.1098/rspb.2019.1910

12. Hamilton, R., Keenan, J. P., Catala, M., & Pascual-Leone, A. (2000). Alexia for Braille following bilateral occipital stroke in an early blind woman. In NeuroReport (Vol. 11, Issue 2, pp. 237–240). Ovid Technologies (Wolters Kluwer Health). https://doi.org/10.1097/00001756-200002070-00003

13. Siuda-Krzywicka, K., Bola, Ł., Paplińska, M., Sumera, E., Jednoróg, K., Marchewka, A., Śliwińska, M. W., Amedi, A., & Szwed, M. (2016). Massive cortical reorganization in sighted Braille readers. In eLife (Vol. 5). eLife Sciences Publications, Ltd. https://doi.org/10.7554/elife.10762

Bola, Ł., Siuda-Krzywicka, K., Paplińska, M., Sumera, E., Zimmermann, M., Jednoróg, K., Marchewka, A., & Szwed, M. (2017). Structural reorganization of the early visual cortex following Braille training in sighted adults. In Scientific Reports (Vol. 7, Issue 1). Springer Science and Business Media LLC. https://doi.org/10.1038/s41598-017-17738-8

14. Merabet, L. B., Hamilton, R., Schlaug, G., Swisher, J. D., Kiriakopoulos, E. T., Pitskel, N. B., Kauffman, T., & Pascual-Leone, A. (2008). Rapid and reversible recruitment of early visual cortex for touch. *PloS one*, 3(8), e3046. https://doi.org/10.1371/journal.pone.0003046

   Interestingly, this research team included both doctors Pascual-Leone and Hamilton who wrote that much-quoted piece on the metamodal brain!

15. Elbert, T., Sterr, A., Rockstroh, B., Pantev, C., Müller, M. M., & Taub, E. (2002). Expansion of the tonotopic area in the auditory cortex of the blind. The Journal of neuroscience : the official journal of the Society for Neuroscience, 22(22), 9941–9944. https://doi.org/10.1523/JNEUROSCI.22-22-09941.2002

16. Sterr, A., Müller, M. M., Elbert, T., Rockstroh, B., Pantev, C., & Taub, E. (1998). Perceptual Correlates of Changes in Cortical Representation of Fingers in Blind Multifinger Braille Readers. In The Journal of Neuroscience (Vol. 18, Issue 11, pp. 4417–4423). Society for Neuroscience. https://doi.org/10.1523/jneurosci.18-11-04417.1998

17. Dietrich, S., Hertrich, I., & Ackermann, H. (2015). Network Modeling for Functional Magnetic Resonance Imaging (fMRI) Signals during Ultra-Fast Speech Comprehension in Late-Blind Listeners. PloS one, 10(7), e0132196. https://doi.org/10.1371/journal.pone.0132196

18. Voss, P., Tabry, V., & Zatorre, R. J. (2015). Trade-off in the sound localization abilities of early blind individuals between the horizontal and vertical planes. The Journal of neuroscience : the official journal of the Society for Neuroscience, 35(15), 6051–6056. https://doi.org/10.1523/JNEUROSCI.4544-14.2015

19. Dufour, A., Després, O., & Candas, V. (2005). Enhanced sensitivity to echo cues in blind subjects. Experimental brain research, 165(4), 515–519. https://doi.org/10.1007/s00221-005-2329-3

20. Sorokowska, A., Sorokowski, P., Karwowski, M. et al. Olfactory perception and blindness: a systematic review and meta-analysis. Psychological Research 83, 1595–1611 (2019). https://doi.org/10.1007/s00426-018-1035-2

21. Manescu, S., Chouinard-Leclaire, C., Collignon, O., Lepore, F., & Frasnelli, J. (2020). Enhanced Odorant Localization Abilities in Congenitally Blind but not in Late-Blind Individuals. In Chemical Senses (Vol. 46). Oxford University Press (OUP). https://doi.org/10.1093/chemse/bjaa073

22. Kleemann, A. M., Albrecht, J., Schöpf, V., Haegler, K., Kopietz, R., Hempel, J. M., Linn, J., Flanagin, V. L., Fesl, G., & Wiesmann, M. (2009). Trigeminal perception is necessary to localize odors. In Physiology & Behavior (Vol. 97, Issues 3–4, pp. 401–405). Elsevier BV. https://doi.org/10.1016/j.physbeh.2009.03.013

23. Sensory Substitution. (n.d.). In Encyclopedia of Neuroscience (pp. 3663–3663). Springer Berlin Heidelberg. https://doi.org/10.1007/978-3-540-29678-2_5346

24. Bach-y-Rita, P. (1967). SENSORY PLASTICITY. In Acta Neurologica Scandinavica (Vol. 43, Issue 4, pp. 417–426). Wiley. https://doi.org/10.1111/j.1600-0404.1967.tb05747.x

25. BACH-Y-RITA, P., COLLINS, C. C., SAUNDERS, F. A., WHITE, B., & SCAD-DEN, L. (1969). Vision Substitution by Tactile Image Projection. In Nature (Vol. 221, Issue 5184, pp. 963–964). Springer Science and Business Media LLC. https://doi.org/10.1038/221963a0

26. The Blind Climber Who "Sees" With His Tongue, Discover Magazine Jun 23, 2008   https://www.discovermagazine.com/mind/the-blind-climber-who-sees-with-his-tongue

27. Jicol, C., Lloyd-Esenkaya, T., Proulx, M. J., Lange-Smith, S., Scheller, M., O'Neill, E., & Petrini, K. (2020). Efficiency of Sensory Substitution Devices Alone and in Combination With Self-Motion for Spatial Navigation in Sighted and Visually Impaired. In Frontiers in Psychology (Vol. 11). Frontiers Media SA. https://doi.org/10.3389/fpsyg.2020.01443

28. Guarniero, G. (1974). Experience of Tactile Vision. In Perception (Vol. 3, Issue 1, pp. 101–104). SAGE Publications. https://doi.org/10.1068/p030101

5. SUPER HEARING

1. See, for instance, episodes two ("*Cut Man*") and seven ("*Stick*") of the first season.

2. This pattern is obviously unique to each person, and your brain is used to the way your own ears are shaped. If you artificially (or by some more dramatic event) change the shape of the outer ear, your brain will eventually adjust to the new pattern.

3. The key to remembering this, if you choose to, is to keep in mind that the "base" does *not* go with the "bass."

4. Liberman, M. C., Gao, J., He, D. Z. Z., Wu, X., Jia, S., & Zuo, J. (2002). Prestin is required for electromotility of the outer hair cell and for the cochlear amplifier. In Nature (Vol. 419, Issue 6904, pp. 300–304). Springer Science and Business Media LLC. https://doi.org/10.1038/nature01059

5. The fundamental frequency is defined as the lowest frequency of a periodic sound. Periodic means that it has a regular repeating pattern, as is typical of many sounds, such as that of a plucked guitar string. However, such sounds contain both the fundamental frequency and higher frequencies that are typically multiples of the former. The fundamental frequency determines the perceived pitch of the sound, whereas the complete mix of frequencies completes the picture so that the same tone can be played by a violin and a piano and have the same pitch yet sound different in other respects. This "other" category is sometimes called the *timbre* of the sound.

6. In the real world, there are natural trade-offs between frequency and temporal resolution, having to do with sampling rates and the nature of sound. In practice, the region of the basilar membrane devoted to higher frequencies is less able than lower frequency regions to register the exact frequency, but better able to resolve events over time. The reverse is true for lower frequency regions. Still, I suppose we could squeeze out a little bit more of each!

7. You may be surprised to see that one order of magnitude is represented by twenty "notches" on the scale, and not ten. Sound *intensity* actually increases by

one order of magnitude for every step of ten on the scale. Sound intensity, which is measured in W/m2, and sound *pressure* are related (the former is proportional to the square of the latter, and are linked by particle velocity) and very often confused for one another.

We usually talk about the term *intensity* in relation to sound output at the source, and the sound *pressure* when specifying a measurement in the "field" away from the source. Yes, this is massively confusing.

8. If this were a light source, each doubling of the distance from the source would lead the light intensity to drop to one quarter. This relationship is due to simple geometry and is described by something called the inverse square law. For sound pressure, the relationship is a little more straightforward so that the law becomes simply an inverse distance law, without the square. The intensity of a sound wave is related to its amplitude squared, which is what turns the inverse square law for intensity into an inverse distance law for sound pressure.

9. Their hearing sweet spot is at around 16,000 Hz, a sound frequency that loses 30 dB to the air over 100 meters, and it's not as if these sounds are particularly loud to begin with.

10. For a look at what is currently the world's quietest place, take a look at this anechoic lab environment built by Microsoft: https://news.microsoft.com/stories/building87/audio-lab.php

11. Though in technological applications and signal processing "noise" can absolutely refer to an unwanted intrusion that interferes with the signal.

12. Cooley, B. (2018, April 4). *Is technology driving your pet insane?* CNET. Retrieved July 10, 2022, from https://www.cnet.com/home/smart-home/is-technology-driving-your-pet-insane/

13. See, for instance, Abbas, A. K., & Bassam, R. (2009). Phonocardiography Signal Processing. In Synthesis Lectures on Biomedical Engineering (Vol. 4, Issue 1, pp. 1–194). Springer Science and Business Media LLC. https://doi.org/10.2200/s00187ed1v01y200904bme031

14. Thought I did find a study looking at the impedance of the chest wall and its effect on more traditional heart measurements: Zimmermann, Henrik & Møller, Henrik & Hansen, John & Hammershøi, Dorte. (2010). Assessment of chest impedance in relation to Phonocardiography.

15. Hambling, D. (2019, June 27). The Pentagon has a laser that can identify people from a distance—by their heartbeat. MIT Technology Review. Retrieved July 10, 2022, from https://www.technologyreview.com/2019/06/27/238884/the-pentagon-has-a-laser-that-can-identify-people-from-a-distanceby-their-heartbeat/

16. Saxe, L. Science and the CQT polygraph. Integr. psych. behav. 26, 223–231 (1991). https://doi.org/10.1007/BF02912514

6. A SENSE OF SPACE

1. Expertly summarized by Samuel Perkins Hayes in "Facial vision Or The Sense of Obstacles" Perkins Publications No. 12June, 1935

Major kudos to the library at the University of Rochester and the Stockholm

City Library for helping me access a copy of this rare book through an interli-brary loan.

2. "Facial Vision": The Perception of Obstacles by the Blind. Michael Supa, Milton Cotzin and Karl M. Dallenbach. The American Journal of Psychology Vol. 57, No. 2 (Apr., 1944), pp. 133-183.

3. I first learned that this movie existed when watching a video presentation by Lore Thaler, one of the most prominent researchers in the human echolocation field. See it here: https://vlp.mpiwg-berlin.mpg.de/library/data/lit39549

4. "Facial Vision:" Perception of Obstacles by the Deaf-Blind, Philip Worchel and Karl M. Dallenbach The American Journal of Psychology Vol. 60, No. 4 (Oct., 1947), pp. 502-553

5. Face to Face (1957), by Ved Metha

6. Thaler, L., Wilson, R. C., & Gee, B. K. (2014). Correlation between vividness of visual imagery and echolocation ability in sighted, echo-naïve people. Experimental Brain Research (Vol. 232, Issue 6, pp. 1915–1925). Springer Science and Business Media LLC. https://doi.org/10.1007/s00221-014-3883-3

7. Ekkel, M. R., van Lier, R., & Steenbergen, B. (2016). Learning to echolocate in sighted people: a correlational study on attention, working memory and spatial abilities. Experimental Brain Research (Vol. 235, Issue 3, pp. 809–818). Springer Science and Business Media LLC. https://doi.org/10.1007/s00221-016-4833-z

8. Teng, Santani, Verena R. Sommer, Dimitrios Pantazis, and Aude Oliva. "Hearing Scenes: A Neuromagnetic Signature of Auditory Source and Reverberant Space Separation." eNeuro 4, no. 1 (January/February 2017). doi:10.1523/eneuro.0007-17.2017.

9. Wallmeier Ludwig, Geßele Nikodemus and Wiegrebe Lutz (2013) Echolocation versus echo suppression in humans Proc. R. Soc. B. 2802013142820131428
    http://doi.org/10.1098/rspb.2013.1428

10. Thaler, Lore. "Echolocation May Have Real-life Advantages for Blind People: An Analysis of Survey Data." Frontiers in Physiology (May 2013). doi:10.3389/fphys.2013.00098.

11. For a review on human echolocation generally, see:
    Thaler, L., & Goodale, M. A. (2016). Echolocation in humans: an overview. In WIREs Cognitive Science (Vol. 7, Issue 6, pp. 382–393). Wiley. https://doi.org/10.1002/wcs.1408

12. See: Vercillo, T., Milne, J. L., Gori, M., & Goodale, M. A. (2015). Enhanced auditory spatial localization in blind echolocators. In Neuropsychologia (Vol. 67, pp. 35–40). Elsevier BV. https://doi.org/10.1016/j.neuropsychologia.2014.12.001

13. Daniel Kish, the founder of World Access For The Blind is a well-known and lifelong echolocator who actually appears among the authors of several of the studies into echolocation. He is also easy to identify as a subject in several of them. I also recommend Kuljit Mithra's interview with Kish on ManWithout-Fear.com

14. Ashmead, D. H., Wall, R. S., Eaton, S. B., Ebinger, K. A., Snook-Hill, M.-M., Guth, D. A., & Yang, X. (1998). Echolocation Reconsidered: Using Spatial Variations in the Ambient Sound Field to Guide Locomotion. In Journal of Visual Impair-

ment & Blindness (Vol. 92, Issue 9, pp. 615–632). SAGE Publications. https://doi.org/10.1177/0145482x9809200905

Ashmead, D. H., & Wall, R. S. (1999). Auditory perception of walls via spectral variations in the ambient sound field. Journal of rehabilitation research and development, 36(4), 313–322.

7. THE FORGOTTEN NOSE

1. "Oliver Sacks' most mind-bending experiment" The Telegraph, 13 November 2012

    http://www.telegraph.co.uk/culture/9661347/Oliver-Sacks-most-mind-bending-experiment.html

2. Having no sense of smell, a condition known as anosmia, is actually not uncommon, though the lack of awareness of this issue parallels our general lack of appreciation for the sense of smell. According to the Monell Center, some 6.3 million Americans report being anosmic, though the true number is likely to be higher. Compare this to the estimated 260,000 Americans who are nearly or totally blind, i.e. have just light perception or less, (American Foundation for the Blind, 2004) or the 600,000 who are functionally deaf (Gallaudet University, 2005), and one has to wonder why the disorder was rarely heard of in the days before the Covid-19 pandemic which has since robbed countless people of their sense of smell for varying amounts of time. It is not unreasonable to see olfaction as less essential than vision or hearing, but anosmia is associated with a higher prevalence of depression than what is seen in the general population, and many anosmics dearly miss their sense of smell.

3. Porter, J., Craven, B., Khan, R. M., Chang, S.-J., Kang, I., Judkewitz, B., Volpe, J., Settles, G., & Sobel, N. (2006). Mechanisms of scent-tracking in humans. In Nature Neuroscience (Vol. 10, Issue 1, pp. 27–29). Springer Science and Business Media LLC. https://doi.org/10.1038/nn1819

4. Bushdid, C., Magnasco, M. O., Vosshall, L. B., & Keller, A. (2014). Humans Can Discriminate More than 1 Trillion Olfactory Stimuli. In Science (Vol. 343, Issue 6177, pp. 1370–1372). American Association for the Advancement of Science (AAAS). https://doi.org/10.1126/science.1249168

5. There are plenty of videos online demonstrating the phenomenon of change blindness, and I recommend you look them up if you are not familiar with the concept.

6. Plailly, J., Delon-Martin, C., & Royet, J.-P. (2011). Experience induces functional reorganization in brain regions involved in odor imagery in perfumers. In Human Brain Mapping (Vol. 33, Issue 1, pp. 224–234). Wiley. https://doi.org/10.1002/hbm.21207

7. See:

    Holland, R. W., Hendriks, M., & Aarts, H. (2005). Smells like Clean Spirit: Nonconscious Effects of Scent on Cognition and Behavior. Psychological Science, 16(9), 689–693. http://www.jstor.org/stable/40064295

    and

Mujica-Parodi, L. R., Strey, H. H., Frederick, B., Savoy, R., Cox, D., Botanov, Y., Tolkunov, D., Rubin, D., & Weber, J. (2009). Chemosensory Cues to Conspecific Emotional Stress Activate Amygdala in Humans. In J. Lauwereyns (Ed.), PLoS ONE (Vol. 4, Issue 7, p. e6415). Public Library of Science (PLoS). https://doi.org/10.1371/journal.pone.0006415

8. Frumin, I., Perl, O., Endevelt-Shapira, Y., Eisen, A., Eshel, N., Heller, I., Shemesh, M., Ravia, A., Sela, L., Arzi, A., & Sobel, N. (2015). A social chemosignaling function for human handshaking. In eLife (Vol. 4). eLife Sciences Publications, Ltd. https://doi.org/10.7554/elife.05154

9. Buck, L. And Axel, R. (1991) 'A novel multigene family may encode odorant receptors: A molecular basis for odor recognition', Cell Vol. 65, pp.175-187.

10. Glusman G, Yanai I, Rubin I, Lancet D. 'The complete human olfactory subgenome'. Genome Res. 2001;11:685–702.

    Zozulya S, Echeverri F, Nguyen T. 'The human olfactory receptor repertoire'. Genome Biol. 2001;2:research0018.1–0018.12.

    Niimura Y, Nei M. 'Evolution of olfactory receptor genes in the human genome'. Proc Natl Acad Sci USA. 2003;100:12235–12240.

11. Gilad, Yoav et al. "Loss of Olfactory Receptor Genes Coincides with the Acquisition of Full Trichromatic Vision in Primates." PLoS Biology 2.1 (2004): e5. PMC. Web. 7 June 2015.

12. Matsui, A., Go, Y., & Niimura, Y. (2010). Degeneration of olfactory receptor gene repertories in primates: no direct link to full trichromatic vision. Molecular biology and evolution, 27(5), 1192–1200. https://doi.org/10.1093/molbev/msq003

13. Olender, T., Keydar, I., Pinto, J. M., Tatarskyy, P., Alkelai, A., Chien, M.-S., Fishilevich, S., Restrepo, D., Matsunami, H., Gilad, Y., & Lancet, D. (2016). The human olfactory transcriptome. In BMC Genomics (Vol. 17, Issue 1). Springer Science and Business Media LLC. https://doi.org/10.1186/s12864-016-2960-3

14. Gelstein, S., Yeshurun, Y., Rozenkrantz, L., Shushan, S., Frumin, I., Roth, Y., & Sobel, N. (2011). Human Tears Contain a Chemosignal. In Science (Vol. 331, Issue 6014, pp. 226–230). American Association for the Advancement of Science (AAAS). https://doi.org/10.1126/science.1198331

8. HEAT AND TOUCH

1. After a reference to the ability to sense color was made as a kind of in-joke in *Daredevil* #8 (2014), by Mark Waid and Chris Samnee, I noticed some fans latch on to this. I actually double-checked with Samnee personally at the time, and can assure everyone that this ability has *not* been reinstated!

2. The visual system is even clever enough to be able to factor in different sources of illumination to keep our world coherent, even though ambiguity on this point can have some interesting effects. Do you remember "the dress" from a few years ago? That Internet sensation that had people arguing over whether the dress in question was black and blue or white and gold? That's how much of a difference the brain's priors can make in situations that we might assume to be straight-forward.

3. Passini, R., & Rainville, C. (1992). The Dermo-Optical Perception of Color as an Information Source for Blind Travelers. Perceptual and Motor Skills, 75(3), 995–1010. https://doi.org/10.2466/pms.1992.75.3.995

4. At least one study has actually noted a difference in temperature discrimination between congenitally blind subjects and sighter controls: Slimani, H., Ptito, M., & Kupers, R. (2015). Enhanced heat discrimination in congenital blindness. In Behavioural Brain Research (Vol. 283, pp. 233–237). Elsevier BV. https://doi.org/10.1016/j.bbr.2015.01.037

5. For an in-depth look at this topic, and other alleged forms of opto-dermic perception, see my post "Have I been too hard on Stan Lee? On reading and sensing colors by touch" (July 2021) at *The Other Murdock Papers*.

6. See this video interview with Blastoff Comics: http://www.blastoffcomics.com/2012/01/the-blastoff-video-interview-mark-waid/

7. It should be noted that lead actor Charlie Cox has made references to filming a scene which featured the ability that was later cut. There has also been some discussion among fans concerning a scene in episode 8 of the first season in which Matt touched a set of blueprints, which he also recognizes as such. I personally don't find this particularly odd, since this material may well be texturally recognizable as blueprints and have discernible features. It doesn't seem obvious that what he is doing in that scene would naturally allow any kind of fluent print reading of the type that was introduced in the Silver Age comics, though opinions will naturally differ.

8. You may occasionally see braille written with a capital 'B' as a reference to Louis Braille. However, smaller-case "braille" is becoming increasingly common and is recommended when the topic is the braille code itself, as opposed to its inventor.

9. The Braille Literacy Crisis in America – Facing the Truth, Reversing the Trend, Empowering the Blind - A Report to the Nation by the National Federation of the Blind Jernigan Institute (March 26, 2009) https://nfb.org//sites/default/files/images/nfb/publications/bm/bm09/bm0905/bm090504.htm

10. Silverman, A. M., & Bell, E. C. (2018). The Association between Braille Reading History and Well-being for Blind Adults. Journal of Blindness Innovation and Research, 8(1). Retrieved from https://nfb.org/images/nfb/publications/jbir/jbir18/jbir080103.html. doi: http://dx.doi.org/10.5241/8-141

11. For an in-depth history of braille, see my October 2009 post about braille history at The Other Murdock Papers

12. I can think of exactly one example in the history of the Daredevil comic where Matt's ability to gauge the number of remaining bullets in a gun this way has actually been used in a story. It happens in a scene in *Daredevil* #17 (1964), by Stan Lee and John Romita Sr. in which Daredevil is battling the Masked Marauder. He picks up a gun dropped by one of the Marauder's men to shoot down a nearby blimp.

9. THE PERPLEXING ORIGINS OF THE RADAR SENSE

1. According to Mark Denny, the author of Blip, Ping & Buzz – Making Sense of Radar and Sonar, the word "ping" is actually primarily associated with sonar interfaces, whereas the "blip" is the corresponding term for radar. Was Stan Lee on top of this stuff? Probably not.

2. Wood did stay on as the inker of Bob Powell's pencils in the next issue

3. See, examples in *Daredevil* #52, *Daredevil* #62, and *Daredevil* #64.

4. In the timeline of comic book historians, the end of the "Silver Age" and the beginning of the "Bronze Age" is usually dated to the year 1970 (See for instance: https://comicbookhistorians.com/the-8-ages-of-comic-books/). The Bronze Age, which ends around the mid-1980s, meant a turn to slightly darker stories and "real world" themes pertaining to various social issues of the day. If we are going to map this change in tone to the publication history of *Daredevil*, I would say that the start of Gerry Conway's run as the writer marks a pretty clear dividing line between Silver Age and Bronze Age *Daredevil*. Conway's first issue was *Daredevil* #72, which is cover-dated to January 1971. Gene Colan as the regular artist obviously straddled these eras and offered some needed consistency.

10. MAKING SENSE OF THE RADAR SENSE

1. For the full story, see: Cronin, B. (2006, April 20). *Comic Book Urban Legends Revealed #47!* Comic Book Resources (CBR). Retrieved June 15, 2022, from https://www.cbr.com/comic-book-urban-legends-revealed-47/

2. Miller had also written the stand-alone issue *Daredevil* #219, for which he also did the artwork. He also co-wrote *Daredevil* #226 with Denny O'Neil. *Daredevil* #227 has February 1986 as the cover date which means it was available in stores a couple of months earlier.

3. This and other excerpts from the so-called "OHOTMU" come from the Official Handbook of the Marvel Universe Deluxe Edition #3 (Feb 1985), written by Mark Gruenwald (writer/producer) and Peter Sanderson (writer/researcher). https://viewcomics.me/the-official-handbook-of-the-marvel-universe-deluxe-edition/issue-3/2

4. According to the OHOTMU, Matt's hearing threshold is 7 dB, which is above that of the average person (for our most sensitive frequencies). His sense of smell, meanwhile, allows him to detect odors at a concentration of thirty parts per million. As we learned in chapter seven, the odor threshold varies depending on the substance, and many odors are detectable at *much* lower concentrations.

5. I say almost, because there is also a scene where you see what looks like reflected smoke(!) somehow revealing Elektra's face when the two meet at a formal event. If you've followed along thus far, you may already have the sense that this is *not* how fumes of cigarette smoke behave, or how scent is detected.

11. THERE IS SOMETHING IT IS LIKE TO RADAR-SENSE

1. The title of this chapter is very much a play on the name of American philosopher Thomas Nagel's 1974 paper What Is It Like to Be a Bat?, a rare example of a scholarly work that is widely known outside its field.* The paper largely revolves around the problems of qualia, though Nagel doesn't use the term, talking instead of the "subjective character of experience." He argues that such experiences are central to the very notion of consciousness. As for the question in the title, Nagel also makes the argument that while there is something it is like to be a bat, we as humans – and thus not in possession of a bat brain – cannot fully comprehend or experience the bat's view of the world in any real sense. In fact, Nagel suggests that it is impossible to describe consciousness in the materialist tradition using references only to mental or bodily processes. Even our own conscious experiences are fully known – and knowable – only to ourselves.

    Nagel's paper is interesting to the Daredevil fan, perhaps not so much for its central claim as for the example chosen: the bat. What makes the echolocating bat interesting to explore from a philosophical perspective is the unique "view" conferred by its ability to locate prey by sound. The bat is thus able to use hearing for tasks that most other mammals solve by using vision, which presumably gives this "seeing-hearing" a qualia distinct from both vision and hearing.

    What Is It Like to Be a Bat? Thomas Nagel, The Philosophical Review Vol. 83, No. 4 (Oct., 1974), pp. 435-450. This paper even has its own Wikipedia entry. https://en.wikipedia.org/wiki/What_Is_It_Like_to_Be_a_Bat%3F

2. For this example, I direct curious readers to the book's companion website.

3. See, for instance, Anil Seth's TED Talk on consciousness where he mentions the wider use of this metaphor.

4. See the 2012 post The philosophy of Paolo Rivera's radar at The Other Murdock Papers.

    http://www.theothermurdockpapers.com/2012/04/the-philosophy-of-paolo-riveras-radar/

5. Kevin Kobasic was the artist for Daredevil #316. Daredevil #328 had a guest creative team with writing by Gregory Wright, and art by Sergio Cariello.

6. Whitney, D., & Levi, D. M. (2011). Visual crowding: a fundamental limit on conscious perception and object recognition. In Trends in Cognitive Sciences (Vol. 15, Issue 4, pp. 160–168). Elsevier BV. https://doi.org/10.1016/j.tics.2011.02.005

7. The peripheral visual field is also characterized by a reduction in color vision that gets more extreme the further you move away from the center. This is not something we typically notice.

8. Theiss, J. D., Bowen, J. D., & Silver, M. A. (2021). Spatial Attention Enhances Crowded Stimulus Encoding Across Modeled Receptive Fields by Increasing Redundancy of Feature Representations. In Neural Computation (Vol. 34, Issue 1, pp. 190–218). MIT Press - Journals. https://doi.org/10.1162/neco_a_01447

12. THE MISSING SENSE

1. Krabben, K. J., van der Kamp, J., & Mann, D. L. (2018). Fight without sight: The contribution of vision to judo performance. In Psychology of Sport and Exercise (Vol. 37, pp. 157–163). Elsevier BV. https://doi.org/10.1016/j.psychsport.2017.08.004

2. I was quite tickled to get confirmation from *Daredevil* season three showrunner Erik Oleson that the eye chart placement in this scene was very much a deliberate choice!

3. This issue was written by D.G. Chichester using the pseudonym "Alan Smithee."

4. Waid's run actually encompasses two volumes of the comic, 2011-2014, and 2014-2015. Paolo Rivera, and Marcos Martín took turns over the first few arcs. Chris Samnne came onboard with Daredevil #12 (2011) and wound up becoming the main artists, with occasional guest stints by other artists. Samnee was also credited as co-writer for much of his and Waid's second run.

5. Alice Woodside Lynch, a well-known Daredevil fan and braille transcriber had this to say about the scene when I wrote about it at *The Other Murdock Papers*: "Here's [...] what I assume Joe Quesada was trying to put on Matt's phone as somewhat of an inside joke. He must have done a little research, but failed to get the dots aligned properly in the six-dot braille cell to really say what I think he meant to say. The first button at least half visible ends with "ex". Next one shows "–ggy", arguably "Foggy". Next is probably meant to be "jay", with the one below that "bob". Neither of those has the dots quite in the right configuration, but the joke fits since there are numerous references to Jay and Bob because Kevin Smith was writing the book at the time. The last button, the one Matt hits, has "sw" on it. Maybe that was to throw people off, thinking he's calling the Scarlet Witch? Nah, since we know he's calling Natasha. Anyway, that's what it says."

6. See the National Federation of the Blind. https://nfb.org/resources/blindness-statistics

Printed in Great Britain
by Amazon

13082816R00185